INTRODUCTORY PROBABILITY THEORY

PROBABILITY: PURE AND APPLIED

A Series of Textbooks and Reference Books

Editor

MARCEL NEUTS

University of Delaware
Newark, Delaware

Other Volumes in Preparation

INTRODUCTORY PROBABILITY THEORY

JANOS GALAMBOS
Temple University
Philadelphia, Pennsylvania

MARCEL DEKKER, INC.   New York and Basel

Library of Congress Cataloging in Publication Data

Galambos, Janos, [date]
    Introductory probability theory.

    (Probability, pure and applied ; 1)
    Bibliography: p.
    Includes index.
    1. Probabilities. I. Title. II. Series.
QA273.G3127 1984   519.2   84-19859
ISBN 0-8247-7179-6

MARCEL DEKKER, INC.
270 Madison Avenue, New York, New York 10016

Current printing (last digit):
10 9 8 7 6 5 4 3 2 1

PRINTED IN THE UNITED STATES OF AMERICA

Preface

This is a one-semester introductory text on probability theory. In this text, I present probability theory as a branch of the mathematical sciences with its many possibilities of direct scientific, technical, engineering, and business applications. By "applications" I mean more than the mere interest of the expert. I believe that a person with a degree should be intelligent enough to understand the ways in which scientific, industrial, and political decisions are made. The worked-out examples are therefore selected with the triple aim of teaching the subject, of providing general intelligence, and of preparing for other courses related to probability theory, such as statistics, business decisions, reliability theory, and others.

I prove relatively early in the book the close relation between the relative frequency and the probability of an event. This is in strong correlation with my overall attempt to present the two faces of probability theory: namely, that the theory itself is a rigorous mathematical discipline, but that when the results of the theory are applied to a specific practical case, the emphasis should be on the practical side. For example, if one were to insist on mathematical rigor, the normal distribution could not be applied to positive random variables, but examples are provided to show that for practical purposes, it is absolutely correct to assume normality even if the variable is known to be positive. At the same time, I point out the danger of being too liberal when choosing the underlying distribution of a random variable. As a matter of fact, I consider it a major point in my approach that the model building be emphasized with the same weight as that given to any other aspect of probability theory.

This material has been tested over a large number of semesters, and proved to be welcomed by students of different backgrounds. Although the basic concepts of calculus are assumed as a prerequisite in the book, I was careful in the selection of examples and exercises to keep the level of calculus to a minimum. In addition, to make the book accessible to students with different majors and interest, I separated lengthier or sophisticated proofs into separate sections, which can be covered or skipped depending on the class.

My specific recommendation is to cover the first five chapters with selected sections of Chapter 6. This last chapter is special in that it contains unrelated topics with quite unexpected results. These sections do not contain exercises. The first five chapters can conveniently be covered in a one-semester course, in which some proofs, and even examples, can be skipped if the class consists of nonmathematics majors. On the other hand, with good mathematical background, the entire book can be covered, because one can then proceed faster. If it is decided to skip some examples, I recommend being very selective, based purely on the mathematical background of the class. By skipping too many examples, the flavor of the book could be lost entirely.

I enjoyed collecting the material for this book over the years, presenting it to my students on three continents, and organizing it into the present book. I only hope that all who teach from it, and who learn from it, will enjoy it equally. I am proud that Marcel Dekker, Inc., is inaugurating a new series with this book under the editorship of my good friend, Professor Macel Neuts. I am indebted to them for this honor. I am also indebted to the editors and all employees of Marcel Dekker, Inc., who worked on my manuscript at various stages of production. I should also thank my wife, Eva Galambos, for drawing the figures and for her contributions to the solutions of the exercises. Finally, let me express my keen interest in the views of those who use this book; please, do not hesitate to send me your views.

Janos Galambos

Contents

INTRODUCTORY
PROBABILITY
THEORY

1
Basic Concepts

1.1 THE SCOPE AND METHODS OF PROBABILITY THEORY

The aim of the theory of probability is to provide tools to measure uncertainty and to discover the laws that govern such a measurement. In other words, we would like to develop a theory which would help us when making decisions in uncertain situations. For example, a manufacturer would like to know the most likely range of the number of its products that can be sold in the forthcoming year. Similarly, an insurance company has to make a good estimate on the number of accidents in the future (for 6 months or a year ahead) in order to set the premium for a policy that the policyholder pays for in advance.

Before we introduce the basic concepts of the theory of probability, let us look at a very simple probabilistic statement which any one of us might have made without applying any theory. Let us analyze the meaning of the statement "X is an honest person." Such a statement is evidently based on past experience, but it actually refers to the future. As a matter of fact, what we are saying is that "since X was reliable on each occasion in the past, X will very likely remain so in the future." Similarly, when we advise someone that "it is not safe to walk here after dark," we base the advice on past experience, but we direct it to the future. Note, however, that there is a dissimilarity between the two preceding statements (regarding honesty and safety, respectively). When we say that X is honest, our past experience led us to an almost certain judgment for the (uncertain) future. However in the case of the safety of a particular area, we imply only that something dangerous may happen. Therefore, this uncertainty is "measured" by us differently from the case of honesty. We want to quantify these likelihoods in such a manner that a clear distinction should emerge between the situations noted here.

The examples of the preceding paragraph are typical ones that a mathematical theory of probability has to describe. The common factor in the two examples was that experience taught us to make predictions for the future. The longer the experience is, the more reliable the prediction will be. The length of experience is evidently measured by the number of times that information becomes available on the phenomenon being observed.

Let us assign accurate mathematical meaning to the concept of experience. "Experience" combines three basic terms: experiment, event, and frequency. We shall use the word *experiment* for anything we perform, either in real terms or hypothetically, or for those things we set out to watch. For example, in the case of studying the safety of a neighborhood, the experiment is to record the number of violent crimes on a single night. Mathematically, we shall always convert an experiment into a set Ω: a set in which all possible outcomes of the experiment are represented. For our example, Ω can thus be chosen as the set $\{0, 1, 2, ...\}$. Now, advising someone about safety, the actual number of crimes may be irrelevant. Our interest reduces to the simple question of whether there was a crime or not. In general, a question posed in connection with an experiment is called an *event*. The fact that an event is always related to an experiment has the mathematical consequence that an event is either the whole Ω or a part of it; that is, events are subsets of Ω. Returning once more to our example of safety, if our interest is the event A—whether there was a crime during a given night—we could then use the set notation $A = \{1, 2, ...\}$.

Finally, a major part of our concept of experience is that we observe the same phenomenon many times. In other words, we repeat a given experiment Ω a large number n of times. We specify an event A in connection with Ω (ask a question), and at each repetition of Ω our only interest is whether A occurred or not. The number $k = k_A = k_A(n)$ of times when A occurred is called the *frequency* of A. The ratio

$$\frac{k_A(n)}{n}$$

is called the *relative frequency* of A. Our belief that by increasing n our prediction for the future becomes more and more reliable is just the lay person's expression for the mathematical concept of convergence of the relative frequency to a number $P(A)$, which depends on the experiment Ω and the event A only. Assuming that such a convergence is indeed mathematically correct, we can then use this number $P(A)$ to measure the uncertainty of A. The major objective of the theory of probability is to show that the relative frequency $k_A(n)/n$, as n increases indefinitely, does converge, in a sense, to a number $P(A)$, which we call the *probability of* (the event) A.

To see the practical importance of this interpretation of probability, let us look at an insurance problem. When a person buys an insurance policy, a form is filled out, on the basis of which the person is put into a specific category of risk. Let us call it category A. Assume that the policy is for 1 year and that everyone in category A is insured for the same amount. Then the insurance company's interest is in whether this person will have a claim within 1 year. But by the client's having been placed into category A, the person's personality is lost and from the company's point of view, the interest lies in whether a person in category A will have a claim (within 1 year). If the company "knows" the probability $P(A)$ of such a claim, and if the company has n clients in this category, then the number (frequency) $k_A(n)$ of claims from category A is "close to" $nP(A)$; that is, the company can calculate the cost of claims in advance, which enables it to set the premium for a predetermined level of profit.

At first, the concept of probability will seem to be somewhat abstract. We shall define the probability $P(\cdot)$ of events through three basic properties, which we call the *axioms*. From these axioms, with mathematical reasoning, we deduce a number of additional properties of probabilities, leading to some general rules of actually calculating probabilities. In the course of the step-by-step development, we shall arrive, in Chapter 2, at the strong relation between the (so-far abstract) probability and the relative frequency of a given event A. From that point on, it will be justified to estimate $nP(A)$ by $k_A(n)$ (as in the example above).

When we limit our investigation to events, then at each outcome of the experiment, there are only two possibilities: a specific event either occurred or it failed to occur. To permit the occurrence of one of several possibilities at each outcome of an experiment Ω, we enlarge our investigation from events to *random variables*. A random variable is nothing but a function defined on Ω; that is, to each member of Ω, we assign a real number. In probability theory, however, instead of the usual questions of calculus, we study those properties of functions that can be associated with probabilistic questions. For example, to a randomly chosen car that needs repair we can assign the (random value of) time needed for its service. A typical question of the service station could be the probability of this service requiring more time than 20 minutes, say. Over a longer period of time, when a large number of cars are brought in, a more appropriate question is to predict the average service time. Questions such as these are treated in Chapters 3 to 6.

After this short summary, we now turn to the detailed development of probability theory.

1.2 EXPERIMENTS AND EVENTS

As was said in the preceding section, the mathematical representation of an experiment is a set Ω in which there corresponds one element to each outcome of the experiment. It is intentional that we do not try to be more specific about Ω, and we thus permit several ways of choosing it. One reason for this ambiguity is mathematical convenience, but another, more important reason, relates to practical considerations. The following two examples reflect these two reasons.

Example 1.1 Give a set Ω that represents the experiment of rolling a die with distinguishable faces.

Whether the distinction of the faces is made by numbers or by colors, we are free to call one face "1," another "2," then "3," "4," "5," and finally "6." Hence the set $\Omega = \{1, 2, 3, 4, 5, 6\}$ represents well the experiment with which we are concerned. But, of course, it would be equally good to choose Ω as {red, brown, green, white, yellow, orange}, say, even if we are handed a black die with a different number of white dots on its faces. In the latter case, our choice of Ω should be accompanied by a "dictionary" stating that "one dot means red," "two mean brown," and so on. Now, if on another occasion, we roll another die for which the labeling of the faces is different from the previous ones, we can still use a previously chosen Ω, but we have to change the dictionary. Since the dictionary will not be a part of our mathematical arguments, we gained from the ambiguity of the definition of Ω in that our mathematical model represented by Ω does not have to be changed when nonessential features of an experiment are modified (such as changing the labeling method of the faces of a die).   ▲

Example 1.2 Give a set Ω which represents the experiment of recording the age of a randomly chosen American citizen.

Should we have tried to define Ω accurately, we would face several difficulties in solving this simple problem. If we count age in years, Ω should contain nonnegative integers (or their equivalents through a dictionary), but which ones? We ought to know the oldest age, and we ought to make sure that we properly recorded all gaps in ages (someone could be 123 years old, but the next-oldest person could be 119). Besides, our Ω would change on the birthday as well as at the death of some persons. This means that a mathematician planning to perform the simple experiment of our example should be familiar with the entire population, including some personal data, in which case the experiment itself is of little interest. This shows the practical value of our requirement on Ω: Its elements should represent all possible outcomes of the experiment, but no further restriction is made on its members. Hence we can choose Ω as the infinite set $\Omega = \{0, 1, 2, \ldots\}$.   ▲

In the following four examples we choose an Ω for the stated experiment and identify specific events as subsets of the chosen Ω.

Example 1.3 Roll a die whose faces are labeled by the numbers 1 through 6. Let A be the event that the result is an even number. Choose an Ω for this experiment and identify A as a subset of Ω.

We choose $\Omega = \{1, 2, 3, 4, 5, 6\}$. Then $A = \{2, 4, 6\}$.   ▲

Example 1.4 Choose a number from among the integers 1 through 10. Let A be the event that the result is an even number. Select an Ω for this experiment and identify A as a subset of Ω.

We can select $\Omega = \{1, 2, 3, 4, 5, 6, 7, 8, 9, 10\}$. We then have $A = \{2, 4, 6, 8, 10\}$.   ▲

Notice that A had the same words in its definition in Examples 1.3 and 1.4. Yet, as sets, they are different. The combination of these two exercises clearly shows what we have stressed at the definition of an event in the preceding section: An event has a meaning only when the experiment has already been specified. It is also made clear through these examples that although we have freedom in choosing Ω, an event is uniquely determined as a subset of Ω.

The following examples are similar to Examples 1.3 and 1.4. In their solutions we use the mathematical symbols usual for sets.

Example 1.5 Roll a black die and a green die, whose faces are numbered 1 through 6. Give an Ω representing this experiment. If A is the event that both of the numbers obtained are smaller than 4, give A as a subset of Ω.

We first agree that the number on the green die is recorded first. Then the following set clearly represents our experiment: $\Omega = \{(i, j): 1 \leq i \leq 6, 1 \leq j \leq 6\}$. With this Ω, $A = \{(i, j): 1 \leq i < 4, 1 \leq j < 4\}$.   ▲

Example 1.6 A hat contains six cards numbered 1 through 6. Pick two cards, one after another, without putting back the first choice. Choose an Ω for this experiment, and give A as a subset of Ω if A is again the event that both of the numbers obtained are smaller than 4.

We can choose $\Omega = \{(i, j): 1 \leq i \leq 6, 1 \leq j \leq 6, i \neq j\}$, and thus $A = \{(i, j): 1 \leq i < 4, 1 \leq j < 4, i \neq j\}$.   ▲

Since events are sets, we can speak of the set-theoretical operations of taking their unions, intersections, and complements. When we use these operations, we always assume that all events involved are connected to the same experiment; that is, they are all subsets of the same Ω. Recall from set theory that the *union*, in the notation $A \cup B$, of the sets A and B is the set of those elements that belong either to A or to B (or to both). On the other hand, the *intersection* $A \cap B$ of the sets A and B is the set of those elements that are in both A and B. Finally, the *complement* A^c of A is the set of all those elements of Ω that do not belong to A.

These operations can best be visualized by means of *Venn diagrams*, in which a set is represented by the interior of a closed curve. The set Ω is the interior of the rectangle in our Venn diagrams. Figure 1.1 represents $A \cup B$, Figure 1.2 shows A^c, and in Figure 1.3 we marked three sets: $A \cap B^c$, $A \cap B$, and $A^c \cap B$.

The following example illustrates the basic operations just introduced.

Example 1.7   Pick a number from the set $\{1, 2, 3, 4, 5, 6, 7, 8\}$. Let A be the event that an even number is obtained; furthermore, let B be the event that the selected number is larger than 4; and finally, C denotes the event that a number smaller than 6 is obtained. List the elements of the following sets (events): A, A^c, $A \cap B$, $B \cap C$, and $A \cup C$.

We can take the set $\{1, 2, 3, 4, 5, 6, 7, 8\}$ as Ω. Thus $A = \{2, 4, 6, 8\}$, $B = \{5, 6, 7, 8\}$, and $C = \{1, 2, 3, 4, 5\}$, from which we have $A^c = \{1, 3, 5, 7\}$, $A \cap B = \{6, 8\}$, $B \cap C = \{5\}$, and $A \cup C = \{1, 2, 3, 4, 5, 6, 8\}$.   ▲

The following basic rules apply to unions, intersections, and complements. They can be verified either by means of Venn diagrams, or by simply showing that every element of the right-hand side of an equation belongs to the left-hand side, and vice versa. Verification is left to the reader.

Group 1: General rules:
Commutative law: $A \cup B = B \cup A$ and $A \cap B = B \cap A$.
Associative law: $(A \cup B) \cup C = A \cup (B \cup C)$
and $(A \cap B) \cap C = A \cap (B \cap C)$.
Distributive law: $A \cap (B \cup C) = (A \cap B) \cup (A \cap C)$.
Group 2: The special events Ω and the empty set $\varnothing$:
$A \cup \Omega = \Omega$, $A \cup \varnothing = A$, $A \cap \Omega = A$, $A \cap \varnothing = \varnothing$
Group 3: The algebraic definition of the complement:
$A \cup A^c = \Omega$, $A \cap A^c = \varnothing$
Group 4: De Morgan's law:
$(A \cup B)^c = A^c \cap B^c$, $(A \cap B)^c = A^c \cup B^c$
Group 5: The operations with a single event:
$A \cup A = A$, $A \cap A = A$, $(A^c)^c = A$
It should be remarked that not all of these rules are independent; some can

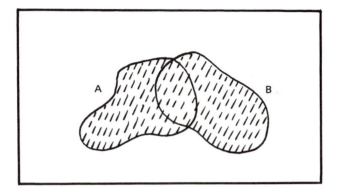

Figure 1.1

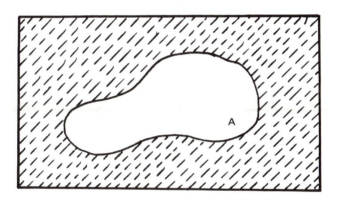

Figure 1.2

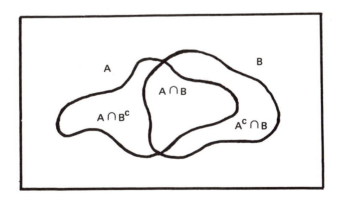

Figure 1.3

be deduced from others. This question, however, is not significant for our purposes.

The rules listed in groups 1 to 5 can be used for proving additional identities among events.

Example 1.8   Show that for arbitrary events A and B

(1.1) $B = (A \cap B) \cup (A^c \cap B)$

and

(1.2) $A \cup B = A \cup (A^c \cap B)$

Although the proof of the foregoing identities is simple by Venn diagrams (see Figures 1.1 and 1.3), we illustrate the application of the basic rules of the operations on events through these identities. Now, to prove (1.1), we start with $B = B \cap \Omega$ from group 2. By the commutative law (group 1) and by the identity $\Omega = A \cup A^c$ from group 3, we thus have $B = (A \cup A^c) \cap B$, which by the commutative and distributive laws of group 1, becomes (1.1), and the proof is complete.

In proving (1.2), the reader is invited to identify the rules from our list which we apply below.

$$A \cup B = (A \cup B) \cap \Omega = (A \cup B) \cap (A \cup A^c)$$
$$= [(A \cup B) \cap A] \cup [(A \cup B) \cap A^c]$$

Now, since A is a subset of $A \cup B$, $(A \cup B) \cap A = A$. Hence

$$A \cup B = A \cup [(A \cup B) \cap A^c]$$
$$= A \cup [(A \cap A^c) \cup (B \cap A^c)]$$
$$= A \cup [\emptyset \cup (A^c \cap B)] = A \cup (A^c \cap B)$$

as stated in (1.2).   ▲

Proofs based on the basic rules of operations with events become more familiar if one considers sets as ordinary numbers, union as addition, and intersection as multiplication. In this reassignment of symbols, Ω becomes unity of multiplication and $\emptyset$ takes the role of zero.

The basic rules listed in groups 1 to 5 can also serve to simplify statements.

Example 1.9   A university board seeks a new president. In the advertisement they state: "For this position, we are looking for a person who is either a noted scientist or who is influential among people in power. Unfortunately, we cannot consider those noted scientists who are not influential among those mentioned." What kind of a person are they seeking?

Let A represent "noted scientist" and B "influential among people in power." Then the president-to-be belongs to $(A \cup B) \cap (A \cap B^c)^c$. By De Morgan's law, this can be simplified to $(A \cup B) \cap (A^c \cup B)$, which, by repeated application of the distributive law, becomes

$$[(A \cup B) \cap A^c] \cup [A \cup B) \cap B]$$
$$= (A \cap A^c) \cup (B \cap A^c) \cup (A \cap B) \cup (B \cap B)$$
$$= \emptyset \cup [B \cap (A^c \cup A)] \cup B = (B \cap \Omega) \cup B$$
$$= B \cup B = B$$

where, in the last steps, we utilized the rules of groups 2, 3, and 5. The result is therefore that the only criterion to be applied in this selection is the ability to influence people in power.   ▲

1.3 THE AXIOMS OF PROBABILITY

Our aim is to introduce the mathematical concept of the probability of an event, which, as indicated in Section 1.1, is expected to describe the asymptotic behavior of relative frequencies. It is therefore natural to look at relative frequencies first and to make sure that whatever property we include in the definition of probability, it does not violate the corresponding property of relative frequencies.

Let us first look at the result of an actual sequence of $n = 40$ rolls of a die. The results were: 3, 2, 1, 1, 5, 2, 1, 5, 4, 3, 4, 4, 2, 5, 5, 2, 4, 6, 3, 4, 6, 2, 3, 6, 1, 2, 4, 3, 3, 4, 2, 2, 5, 1, 6, 5, 3, 6, 4, and 5. If A is the event that the outcome is even, B is that the outcome is 5, and C is that the outcome is either even or 5 (i.e., $C = A \cup B$), then, by simple counting, we get $k_A(40) = 21$, $k_B(40) = 7$, and $k_C(40) = 28$. We can immediately recognize that

$$0 \le k_A(40), \, k_B(40), \, k_C(40) \le 40$$

and

(1.3) $$k_C(40) = k_A(40) + k_B(40)$$

Are these relations specific to A, B, and C and to $n = 40$? Evidently not. By the definition of the frequency of an event A (the specific one above or any other),

$$0 \le k_A(n) \le n$$

or, what this amounts to,

(1.4)
$$0 \leqslant \frac{k_A(n)}{n} \leqslant 1$$

Now, when we look at (1.3) more closely, we can immediately conclude that its validity is due to the fact that $C = A \cup B$ and, at the same time, $A \cap B = \emptyset$. As a matter of fact, whenever we find that C occurred, then either A or B had to occur, but both could not. We thus have the general rule that if $C = A \cup B$ and $A \cap B = \emptyset$, then

$$k_C(n) = k_A(n) + k_B(n)$$

or, after dividing by n,

(1.5) $\dfrac{k_C(n)}{n} = \dfrac{k_A(n)}{n} + \dfrac{k_B(n)}{n}$ where $C = A \cup B, \quad A \cap B = \emptyset$

Finally, let us point out an even more evident fact. In the example of rolling a die, $\Omega = \{1, 2, 3, 4, 5, 6\}$. Thus $k_\Omega(40) = 40$, and, in general, $k_\Omega(n) = n$. In fact, $k_\Omega(n) = n$ whatever the experiment Ω for the simple reason that if Ω failed just once [i.e., $k_\Omega(n) < n$], something occurred at this time which is not represented in Ω, contradicting our choice of Ω. Consequently,

(1.6)
$$\frac{k_\Omega(n)}{n} = 1 \qquad \text{for all } n \geqslant 1$$

Since (1.4) to (1.6) are valid for every n, similar properties should be valid for their limits (if they exist). That is, if probability is indeed the limit of the relative frequency in some sense, its properties should be in agreement with those indicated by (1.4) to (1.6). With this in mind as a guide, we now introduce the following definition.

Definition 1.1   To every event A in connection with a random experiment Ω, we assign a real number $P(A)$, called the *probability* of A, which satisfies the following three properties:
(A1) $0 \leqslant P(A) \leqslant 1$
(A2) $P(\Omega) = 1$
(A3) If $A_1, A_2, \ldots$ is a finite or countably infinite sequence of events such that, for every $i \neq j$, $A_i \cap A_j = \emptyset$, then

$$P(A_1 \cup A_2 \cup \cdots) = P(A_1) + P(A_2) + \cdots$$

We shall refer to the three properties (A1) to (A3) as the *axioms* of probability. Recall that a set is called countably infinite if it can be arranged into a sequence (i.e., we can identify one of them as the first, another one as the second, etc.).

For easier reference in the future, we call two events A and B *mutually exclusive* if $A \cap B = \emptyset$. It is important to remember that $(A3)$ permits us to calculate the probability of a union as the sum of the individual probabilities only if all pairs of the events in question are mutually exclusive.

Notice that the axioms do not instruct us as to how to calculate probabilities. However, in the forthcoming sections, we shall see that a large variety of problems can be solved by simple mathematical deduction from the axioms.

So far, we have defined events as subsets of Ω. Definition 1.1 raises the mathematical question: Given a set Ω, can one assign $P(A)$ to all subsets of Ω such that $(A1)$ to $(A3)$ should hold? It turns out that such $P(A)$ can sometimes be defined only on some collection of subsets of Ω but not on the set of all of its subsets. Although we do not deal with this question in the present book, in the following definition the basic properties of the collection of events are given.

Definition 1.2   For a given Ω, an event is a subset of Ω. The collection $\mathcal{A}$ of all events satisfies the following conditions:
(i)   If $A \in \mathcal{A}$, then $A^c \in \mathcal{A}$.
(ii)   If $A_1, A_2, \ldots$ is a finite or countably infinite sequence such that for each i, $A_i \in \mathcal{A}$, then both

$$A_1 \cup A_2 \cup \cdots \in \mathcal{A}$$

and

$$A_1 \cap A_2 \cap \cdots \in \mathcal{A}$$

(iii) $\Omega \in \mathcal{A}$.

Notice that it was not necessary to include both unions and intersections in (ii). One follows from the other through De Morgan's law because of (i). It should also be emphasized that in view of (i) and (iii), $\emptyset \in \mathcal{A}$.

We use Definition 1.2 only to the extent that we freely use the operations of union, intersection, and complement, whether $\mathcal{A}$ contains all or only some subsets of Ω.

With these definitions at hand, particularly Definition 1.1, we are ready to start the development of the theory of probability. We now conclude this section with the introduction of some further names of concepts specified previously. Ω is referred to in several ways in the literature. We identified it with the experiment itself. It is also called *sample space*, and, in view of $(A2)$, the *sure event*. Since $\emptyset = \Omega^c$, $\emptyset$ is called both the *empty set* and the *impossible event*. The triplet $(\Omega, \mathcal{A}, P)$ is called the *probability space*. Axioms $(A1)$ to $(A3)$ are known as the *Kolmogorov axioms of probability*, because the theory that we are developing was founded by the Soviet mathematician A. N. Kolmogorov in 1933. The classical theory (and thus the major part of Chapters 1 and 2) of course goes back several centuries in the mathematical literature (although scattered in different branches of mathematics, and not always welcomed by other mathematicians, owing mainly to its lack of proper foundation).

1.4   ELEMENTARY CONSEQUENCES OF THE AXIOMS

All statements in the present section are true for every probability space (Ω, $\mathcal{A}$, P).

Theorem 1.1 $P(A^c) = 1 - P(A)$. In particular, $P(\emptyset) = 0$.

Proof: We recall the algebraic definition of the complement (group 3 of the rules in Section 1.2):

$$A \cup A^c = \Omega \qquad \text{and} \qquad A \cap A^c = \emptyset$$

The second equation implies that ($A3$) is applicable to the first one. We get

$$P(\Omega) = P(A \cup A^c) = P(A) + P(A^c)$$

which is the relation that is claimed in view of ($A2$). The particular case of $P(\emptyset) = 0$ follows immediately from the first part of the theorem upon observing that $\emptyset = \Omega^c$ and by one more appeal to ($A2$).   ▲

Theorem 1.2 $P(A^c \cap B) = P(B) - P(A \cap B)$.

Proof: We established in (1.1) the relation

$$B = (A \cap B) \cup (A^c \cap B)$$

Since the events $A \cap B$ and $A^c \cap B$ are mutually exclusive (the first one is a part of A, while the second is a part of A^c), we can again apply ($A3$). Hence

$$P(B) = P(A \cap B) + P(A^c \cap B)$$

which is what was to be proved.   ▲

Theorem 1.3   If $A \subset B$, then $P(A^c \cap B) = P(B) - P(A)$. Consequently, $P(A) \leqslant P(B)$ whenever $A \subset B$.

Proof: The first part follows immediately from Theorem 1.2, because $A \subset B$ evidently implies that $A \cap B = A$. Now, continuing with the equation just proved, ($A1$) yields that the left-hand side, and thus the right-hand side as well, is nonnegative; that is, $P(A) \leqslant P(B)$, as stated. The proof is complete.   ▲

Theorem 1.4 $P(A \cup B) = P(A) + P(B) - P(A \cap B)$.

Proof: We have seen in (1.2) that

$$A \cup B = A \cup (A^c \cap B)$$

Although the right-hand side looks more complicated, its advantage is that the terms in the union are mutually exclusive (the second term is a part of the complement of A); therefore, $(A3)$ is applicable to the right-hand side. We get

$$P(A \cup B) = P(A) + P(A^c \cap B)$$

Theorem 1.2 now completes the proof.   ▲

Compare Theorem 1.4 with $(A3)$. If we assume further that A and B are mutually exclusive, then the last term in Theorem 1.4 is zero (see Theorem 1.1). Consequently, it looks as if we proved $(A3)$ for two events as a special case of Theorem 1.4. This is not the case, however, since we appealed to $(A3)$ in the proof of Theorem 1.4.

Notice that the combined meaning of Theorems 1.1 to 1.4 is this: Given two events A and B, then out of the 12 possible probabilities that we may face—$P(A)$, $P(B)$, $P(A^c)$, $P(B^c)$, $P(A \cup B)$, $P(A^c \cup B)$, $P(A \cup B^c)$, $P(A^c \cup B^c)$, $P(A \cap B)$, $P(A^c \cap B)$, $P(A \cap B^c)$, and $P(A^c \cap B^c)$—at most three can be "chosen"; the others are determined uniquely by these 12.

Example 1.10   Assume that $P(A) = 0.7$, $P(B) = 0.5$, and $P(A \cap B) = 0.3$. Evaluate the other nine probabilities listed above.

First, by Theorem 1.1, $P(A^c) = 1 - 0.7 = 0.3$ and $P(B^c) = 1 - 0.5 = 0.5$. Next, from Theorem 1.2 we get $P(A^c \cap B) = 0.5 - 0.3 = 0.2$, $P(A \cap B^c) = 0.7 - 0.3 = 0.4$, and $P(A^c \cap B^c) = P(B^c) - P(A \cap B^c) = 0.5 - 0.4 = 0.1$. Finally, the formula of Theorem 1.4 yields $P(A \cup B) = 0.7 + 0.5 - 0.3 = 0.9$, $P(A^c \cup B) = 0.3 + 0.5 - 0.2 = 0.6$, $P(A \cup B^c) = 0.7 + 0.5 - 0.4 = 0.8$, and $P(A^c \cup B^c) = 0.3 + 0.5 - 0.1 = 0.7$. (Notice that the last four values could be obtained from the preceding four by De Morgan's law. How?)   ▲

Theorem 1.3 guards us against "wrong choices" of probabilities.

Example 1.11   Why is it incorrect to assume that for some events A and B, $P(A) = 0.4$, $P(B) = 0.3$, and $P(A \cap B) = 0.35$?

For arbitrary events, $A \cap B \subset B$ (or A), and thus, by Theorem 1.3, $P(A \cap B) \leq P(B)$, which is not satisfied by the values of the example. ▲

Let us return to establishing some additional general results on probabilities.

Theorem 1.5   For any $n > 1$, and for an arbitrary sequence $A_1, A_2, \ldots$ of events,

$$P(A_1 \cup A_2 \cup \cdots \cup A_n) \leqslant P(A_1) + P(A_2) + \cdots + P(A_n)$$

Proof: Let us first prove the theorem for $n = 2$. That is,

(1.7) $$P(A_1 \cup A_2) \leqslant P(A_1) + P(A_2)$$

This is indeed true. As a matter of fact, by Theorem 1.4,

$$P(A_1 \cup A_2) = P(A_1) + P(A_2) - P(A_1 \cap A_2)$$

and since every probability is nonnegative, the right-hand side is increased if we drop its last term.

Now, writing

$$\bigcup_{j=1}^{n} A_j = \left(\bigcup_{j=1}^{n-1} A_j \right) \cup A_n$$

and considering the right-hand side as the union of just two events, we get from (1.7),

(1.8) $$P\left(\bigcup_{j=1}^{n} A_j \right) \leqslant P\left(\bigcup_{j=1}^{n-1} A_j \right) + P(A_n)$$

This inequality, by the following argument, known as the *method of induction*, leads to Theorem 1.5. Plug into (1.8) the value $n = 3$. Then, (1.8), combined with (1.7), yields the inequality claimed for $n = 3$. Next, substitute n by 4 in (1.8). Then $n - 1 = 3$, for which we have just proved the theorem, which, when applied in (1.8), gives the theorem for $n = 4$. Continuing in this manner, step by step we get the theorem for larger and larger values of n. That is, every value of n will be reached, which completes the proof.   ▲

Example 1.12   Consider a piece of equipment that has 25 components. Assume that each component has probability 0.998 of properly functioning during a trial period. Estimate the probability that all components function properly during this trial period.

Let us mark the components by the numbers 1 through 25. Let A_j be the event that the jth component fails, and B be the event that all components function properly during the trial period. Then

$$B^c = A_1 \cup A_2 \cup \cdots \cup A_{25}$$

and thus, by Theorems 1.1 and 1.5,

$$P(B) = 1 - P(B^c) = 1 - P\left(\bigcup_{j=1}^{25} A_j\right) \geq 1 - \sum_{j=1}^{25} P(A_j)$$
$$= 1 - 25(1 - 0.998) = 0.95$$

Notice that this estimate is valid whatever the construction of the equipment. That is, the interrelation of the components affects only intersections (which represent concurrent functoning or failure). Assuming that probability is indeed the asymptotic value of relative frequencies, the estimate above tells us that fewer than 5% of all pieces of equipment will fail during the trial period. Hence, by adding 5% of the production cost to the retail price, a "free warranty" can be offered without cost to the manufacturer.   ▲

1.5 CLASSICAL PROBABILITY SPACES

The first mathematical questions in connection with "random experiments" were posed by gamblers of seventeenth-century France. In a casino, one rolls dice, bets on roulette, plays cards, and so on. Common to all these is that there are only a finite number of possible outcomes, and all basic outcomes are supposed to be equally likely. When these characteristics are put into a mathematical model, we arrive at the classical probability spaces. We shall see that these models have a much wider field of applicability than just casinos.

Definition 1.3 If the set $\Omega = \{a_1, a_2, \ldots, a_N\}$ of all possible outcomes is finite, and if all outcomes are equally likely [i.e., $P(\{a_j\}) = p$ does not depend on j], then the probability space $(\Omega, \mathcal{A}, P)$ is called a *classical probability space*. Here $\mathcal{A}$ is the set of all subsets of Ω. In other words, every subset of Ω is an event.

Let us first deduce from the axioms that the common value

(1.9) $$p = P(\{a_j\}) = \frac{1}{N}$$

Let us set $A_j = \{a_j\}, 1 \leq j \leq N$. Then the events $A_1, A_2, \ldots, A_N$ are mutually exclusive and $A_1 \cup A_2 \cup \cdots \cup A_N = \Omega$. Hence, by the definition of p, and by axioms $(A2)$ and $(A3)$,

$$1 = P(\Omega) = P\left(\bigcup_{j=1}^{N} A_j\right) = \sum_{j=1}^{N} P(A_j) = Np$$

which is equivalent to (1.9).

What is the practical meaning of the assumption that "every outcome is equally likely"? This is just an assumption in several cases of practice as well. When we roll dice in a casino, we trust (assume) that each face will turn up with the same likelihood; when we play cards, "well shuffled" is the equivalent of the assumption that cards are dealt with equal probabilities. Selecting a representative sample from a population is also a classical type of problem; to be scientific in such selections, tables of random numbers can be utilized, but in most cases less scientific choice of individuals will do; in such cases we again believe that no bias was present in the selection.

The mathematical consequence of the assumption that all outcomes are equally likely is formula (1.9) and the fact that Ω now has to be chosen with care. It has to have *exactly* as many elements as outcomes that are possible for the experiment.

Let us interpret (1.9) for some concrete "experiments." When we roll a regular (unbiased) die, $N = 6$, and we have thus *proved* that each face should turn up with probability $p = 1/6$. Similarly, from a well-shuffled deck of 52 cards, any particular card will be dealt with probability $p = 1/52$. Now, select two cards from a well-shuffled deck (of 52). What is the probability that any particular pair is obtained? By (1.9), the answer is $p = 1/N$, where N is the number of ways of selecting 2 out of 52 (different) cards. We can count this N as follows. Place two boxes beside each other. Put one card into the first box; this can be done in 52 ways. Then put one of the remaining cards into the second box; this can be done in 51 ways. Since to every first choice, there are 51 second choices, and there are 52 first choices, we got $52 \times 51 = 2652$ pairs in which we distinguish the members that are first and those that are second. However, if our interest is only in the actual two cards obtained and not the order in which they were dealt, then above, we counted every pair twice. Hence $N = 2652/2 = 1326$ and $p = 1/1326$.

In the counting technique that we just followed, the number 52 can evidently be replaced by any number, and the extension to selecting more than two is straightforward as well. Let us therefore record the preceding argument in general terms.

Selection Without Replacement: Let a lot contain T distinct items. Assume that t items are selected from this lot, one by one (and thus the order of the selected items is relevant) and without the replacement of items selected previously. The number of ways of carrying out this selection procedure is

$$T(T - 1)(T - 2) \cdots (T - t + 1)$$

Notice that when $T = t$, the selection method just described can also be considered as rearranging the order of t items. Namely, imagine the t items in a sequence, labeled 1 through t. Then, selecting all t items, one by one, we actually place one of the items into the first position, another into the

second, and so on, until the remaining item is put into the tth position. A rearrangement of the order of items is called a *permutation*. We thus got as a special case that the number of permutations of t items is $t(t-1)(t-2)\cdots 3\times 2\times 1$, which we abbreviate as $t!$, read t *factorial*. (The factorials increase extremely fast as a function of t: $2!=2$, $3!=6$, $4!=24$, $5!=120$, $6!=720$, $7!=5040$, $8!=40{,}320$, $9!=362{,}880$, $10!=3{,}628{,}800$, etc.)

When we combine the model of selection without replacement with the formula for the number of permutations, we get the following combinatorial result.

Combinations: Select t out of a lot of T distinct items, where repetitions are not permitted and the order of selection is irrelevant. This is called a *combination* of t of the T items. The number of such combinations equals

$$(1.10) \qquad \frac{T(T-1)(T-2)\cdots (T-t+1)}{t!} \qquad 1\le t\le T$$

As a matter of fact, from every combination of t of the T items we can get as many ordered selections of t as there are permutations of t items, which has just been shown to be $t!$. Hence, if the number of combinations of t of the T items is denoted by $\binom{T}{t}$, then $t!\binom{T}{t}$ is the number of possible selections of t without replacement from a lot of T items. That is,

$$t!\binom{T}{t} = T(T-1)(T-2)\cdots (T-t+1)$$

which is just another form of (1.10). The ratio in (1.10) is also known as the *binomial coefficient* (with parameters T and t), and we shall use the notation $\binom{T}{t}$ for it. Notice that if we multiply both the numerator and the denominator of (1.10) by $(T-t)!$, the binomial coefficient $\binom{T}{t}$ takes the following form ($0\le t\le T$, where, by convention, $0!=1$):

$$(1.11) \qquad \binom{T}{t} = \frac{T(T-1)(T-2)\cdots (T-t+1)}{t!} = \frac{T!}{t!(T-t)!}$$

from which it is immediate that

$$(1.12) \qquad \binom{T}{t} = \binom{T}{T-t}$$

Let us now return to calculating probabilities. We adopt the convention that when selection is from a finite set, "at random" refers to "equally likely outcomes."

Example 1.13a   From the numbers 1 through 10, select three at random, one by one and without replacement. Find the probability that the numbers 1, 2, and 3 (in this order) are obtained.

This selection method conforms to the model of selection without replacement. According to its formula, there are $N = 10 \times 9 \times 8 = 720$ possible outcomes of this experiment. From (1.9), the probability of any particular outcome therefore equals $p = 1/720$.   ▲

Example 1.13b   From the numbers 1 through 10, select three at random. Find the probability that the numbers 1, 2, and 3 are obtained.

This time, we were not requested to identify a first, a second, and a third, and thus the number of possible outcomes is the number of combinations of 3 of the 10 numbers. By (1.10), it is equal to $10 \times 9 \times 8/3! = 120$, and thus the probability of getting the numbers 1, 2, and 3 is $p = 1/120$.   ▲

Example 1.14   Four friends, called by their first initials, A, B, C, and D, are seated at random in a row. What is the probability that they sit in the order $ABCD$?

Since they are seated at random in a row, any permutation of the four names is possible as their order of sitting. Hence the number of outcomes of this "experiment" is $N = 4! = 24$, which, in view of (1.9), yields that the probability of any particular order of their seats is $p = 1/24$.   ▲

Example 1.15   In Example 1.14 find the probability that A and B are seated adjacent to each other.

In Example 1.14 we saw that there are $N = 24$ outcomes in the experiment of seating four persons at random in a row; in notational form,

$$\Omega = \{a_1, a_2, ..., a_{24}\}$$

where each a_j is a permutation of the names A, B, C, and D. Let E be the event that A and B are seated adjacent to each other. Then $E \subset \Omega$, and we are free to label the outcomes so that

$$E = \{a_1, a_2, ..., a_k\}$$

with some k. In fact, we can easily list all the a_j that are favorable to E. These are: $a_1 = ABCD$, $a_2 = BACD$, $a_3 = ABDC$, $a_4 = BADC$, $a_5 = CABD$, $a_6 = CBAD$, $a_7 = DABC$, $a_8 = DBAC$, $a_9 = CDAB$, $a_{10} = CDBA$,

$a_{11} = DCAB$, and $a_{12} = DCBA$. That is, we find $k = 12$. Now, we can also write

$$E = \{a_1\} \cup \{a_2\} \cup \cdots \cup \{a_{12}\}$$

where the $\{a_j\}$ evidently are mutually exclusive. Hence, by axiom $(A3)$,

$$P(E) = P(\{a_1\}) + P(\{a_2\}) + \cdots + P(\{a_{12}\})$$

Now, in Example 1.14 we saw that $P(\{a_j\}) = 1/24$ for each j, and thus $P(E) = 12/24 = 1/2$.   ▲

The argument above can be followed in order to evaluate the probability of an arbitrary event E in a classical model. Let us therefore repeat this argument in general terms.

Let $\Omega = \{a_1, a_2, \ldots, a_N\}$ and let E be an arbitrary event. Then $E \subset \Omega$; that is, with some subscripts $1 \leqslant i_1 < i_2 < \cdots < i_k \leqslant N$,

$$E = \{a_{i_1}, a_{i_2}, \ldots, a_{i_k}\}$$

or

$$E = \{a_{i_1}\} \cup \{a_{i_2}\} \cup \cdots \cup \{a_{i_k}\}$$

By an appeal to axiom $(A3)$ and to (1.9), we get

$$P(E) = \sum_{j=1}^{k} P(\{a_{i_j}\}) = k\left(\frac{1}{N}\right) = \frac{k}{N}$$

The significance of this formula is that, in a classical probability space, it is irrelevant which elements of Ω belong to E; only their number k matters. We summarize this very important result as a theorem.

Theorem 1.6   If, in a classical probability space, the number of possible outcomes is N, and if the number of favorable outcomes to the event E is k (in other words, the number of elements in E as a subset of Ω is k), then

$$P(E) = \frac{k}{N}$$

Theorem 1.6 reduces the calculation of probabilities in classical spaces to counting the number of elements of certain sets. The reader should therefore develop as a routine the following basic rule of counting, which is frequently applied in what follows without actually referring to it: If the number of elements of a set can be counted in two steps so that there are n possibilities in the first step which lead to elements of the set, and to every possibility in the first step there are m possibilities in the second step, then the number of elements in the set is nm.

The proof of this rule is simple. Call a possibility in the first step a_i, $1 \leq i \leq n$, and in the second step b_j, $1 \leq j \leq m$. Then a typical element of the set is (a_i, b_j). Now, let us arrange these pairs in a rectangle so that we place (a_i, b_j) in the ith row and the jth column. By assumption, there are then n rows, and in every row there are m entries. So every entry of the $n \times m$ rectangle is filled in; that is, it has nm elements, as claimed.

This rule of counting evidently extends to any number of steps: If the counting of the number of elements of a set can be split into s steps so that there are n_1 possibilities in the first step which lead to elements of the set, and to every first possibility, there are n_2 possibilities in the second step, furthermore, to every pair of possibilities in the first two steps, there are n_3 possibilities in the third step, and so on, then the number of elements of the set is $n_1 \, n_2 \ldots n_s$.

The proof can be done by induction over s from the case of $s = 2$, and its details are thus omitted.

As an example, let us count the number of those three-digit integers whose digits are all distinct. This can be split into three steps by first counting the possible choices of the first digit (this number is nine—any of the integers 1 through 9), then to every choice of the first digit, the number of possibilities for the second digit (nine again—any of the integers 0 through 9 except the value of the first digit), and finally, to every choice of the first two digits, the number of possible values of the third digit (it is eight; why?). By the rule above we thus get that there are $9 \times 9 \times 8 = 648$ three-digit numbers with distinct digits.

Example 1.16   Roll two regular dice. What is the probability that the sum of the numbers on the two faces is 7?

Let us call the numbers on the two faces x and y. Then there are six possibilities for x, and to each value of x there are six possibilities for y, so $N = 6 \times 6 = 36$. Now, if E is the event that $x + y = 7$, then the number k of favorable cases to E can be enumerated by simply listing these cases: $(1, 6)$, $(2, 5), (3, 4), (4, 3), (5, 2)$, and $(6, 1)$, where the first number signifies x, say. We find that $k = 6$, and thus, by Theorem 1.6, $P(E) = 6/36 = 1/6$.   ▲

Example 1.17   Pick a number x at random from the integers 1 through 120. Determine the probability that x is divisible either by 4 or by 6.

Let A represent the event that x is divisible by 4 and B the event that x is divisible by 6. Then we are to determine $P(A \cup B)$. We know from Theorem 1.4 that

$$P(A \cup B) = P(A) + P(B) - P(A \cap B)$$

Now, $P(A) = 30/120$, because there are $N = 120$ possibilities for x, and

$k = 30$ of these are favorable to E, that is, each integer multiple of 4 up to 120. Similarly, $P(B) = 20/120$, because the number of integer multiples of 6 up to 120 equals 20. Finally, since $A \cap B$ means that x is divisible by both 4 and 6, $A \cap B$ is the set of all integers up to 120 which is divisible by 12, and thus $P(A \cap B) = 10/120$. Upon combining these numbers, the quoted formula yields

$$P(A \cup B) = \frac{30}{120} + \frac{20}{120} - \frac{10}{120} = \frac{40}{120} = \frac{1}{3} \qquad \blacktriangle$$

Example 1.18   Roll three regular dice. Find the probability that at least one even number is obtained.

The problem looks very similar to Example 1.17, and indeed we can proceed to solve the problem as the probability of a union. A simpler solution is obtained, however, if we first compute the probability of the complement of the event in the exercise. As a matter of fact, if we let A represent the event that at least one even number is obtained, then $A^c = \{$each face shows an odd number$\}$. Hence, in rolling three dice, the number of favorable cases of A^c is $k = 3 \times 3 \times 3 = 27$, because on each die there are three odd values. On the other hand, the number of all possibilities in this experiment is $N = 6 \times 6 \times 6 = 216$, and thus $P(A^c) = k/N = 27/216 = 1/8$. Consequently (see Theorem 1.1), $P(A) = 1 - 1/8 = 7/8$. $\blacktriangle$

In the next several sections, particular classical probability spaces are discussed.

1.6   SELECTION WITHOUT REPLACEMENT: THE CASE OF TWO TYPES OF ITEMS

This section is devoted to the following selection problem.

A lot contains T items, each of which belongs to one of two types, type I and type II. Let M denote the number of type I items. We select t items from the lot, one by one, without replacing the items already chosen. The selection is done in such a way that each subset of t items from the total lot (i.e., each combination of t of the total number T) has the same probability of being selected. Let X be the number of type I items among the t items selected.

Here X is a random value, that is, a value that depends on the actual outcome of the (random) selection procedure adopted. What can be said about X? Which are its most likely values? How can the ultimately known value of X (at the completion of the selections) be utilized to get information on T or M? We shall concentrate on these questions, but first let us formulate a few practical examples for our selection model.

Example 1.19 (Quality Control) At a manufacturing plant, T products are produced, out of which M are defective. The value of T is usually known, but M is not. The management of the factory (or the buyer) would like to know the most likely range of M on the base of inspecting some (t), among which m defective products are found.   ▲

Example 1.20 (Opinion Polls) In a population of size T, M persons are in favor of a proposal; the rest are against it (let us ignore for the moment the fact that some people could be indifferent). The interest of the pollsters is to give a good approximation to M through selecting t members for interview, among whom m persons turn out to be in favor of the proposal.   ▲

Example 1.21 (Capture-Recapture Model) We would like to know the size T of a wild animal population. For this purpose, M animals are captured and tagged and then set free. After waiting for a period of time, sufficient for the tagged animals to mix with their original population, t are captured again and the number of tagged ones are counted (m). One would like to get reliable information on T, based on the values of M, t, and m.   ▲

It is clear that the selection method of each of the three examples is "well represented" by our selection model formulated at the beginning of this section. Therefore, conclusions within that model are readily applicable to the corresponding questions concerning any one of Examples 1.19 to 1.21. Notice the difference in attitude between the abstract model that we investigate here and the practitioner's problems in the examples. The mathematician (in the abstract model) views the values T, M, and t as fixed, and without performing the selections, tries to say something about the outcome (the number X of type I items). To the mathematician, X is the only variable. The practitioner, on the other hand, did complete the selections, has three numbers at hand (T, t, and m in Examples 1.19 and 1.20 and M, t, and m in Example 1.21), and needs information about the missing one (M or T). The most significant difference is that the number m of type I items among the items selected is invariably known to the practitioner. Now, a mathematical result becomes a practical one by adding: "If the practical value of X is m then this and this relation holds." This difference in emphasis is the essence of mathematical modeling, and it is not restricted to this single selection model.

We now give the fundamental result for the selection model of the present section.

Theorem 1.7 For the selection model described at the beginning of this section, let B_m represent the event that $X = m$. Then

$$(1.13) \qquad p_m = P(B_m) = \frac{\binom{M}{m}\binom{T-M}{t-m}}{\binom{T}{t}}$$

Proof: Because of the assumption that each combination of t of the T items has the same probability, our model becomes a classical one if we take these combinations as the basic outcomes of the experiment. Hence $N = \binom{T}{t}$, the number of the combinations of t of the T items [see (1.10) and (1.11)]. In counting the number of favorable cases to B_m, we note that a combination of t of the T items is favorable to B_m if, and only if, it consists of m type I items, and the remaining $t - m$ items are type II. Now, by (1.10) again, we can select m out of the total M type I items in $\binom{M}{m}$ ways, and to each combination of m of the M type I items, the $t - m$ type II items can be selected in $\binom{T-M}{t-m}$ ways. Consequently, the number k of favorable cases to B_m is the product $k = \binom{M}{m}\binom{T-M}{t-m}$. A substitution into the formula of Theorem 1.6 now leads to (1.13), which completes the proof.   ▲

We shall refer to the probabilities in (1.13) as the *hypergeometric probabilities*. When necessary, we add a phrase and call these the hypergeometric probabilities with parameters T, M, and t. The name comes from another branch of mathematics, where these same numbers appear. The number m is a variable, and it takes the nonnegative integers $0, 1, 2, \ldots$. Evidently, $m \leq M$ and $m \leq t$, and thus the last possible value of m is min (M, t). For brevity, let us put $m^* = \min (M, t)$. Then

$$B_0 \cup B_1 \cup \cdots \cup B_{m^*} = \Omega$$

and the B_j are mutually exclusive. Therefore, by the axioms,

$$1 = P(\Omega) = P\left(\bigcup_{j=1}^{m^*} B_j\right) = \sum_{j=1}^{m^*} P(B_j) = \sum_{j=1}^{m^*} p_j$$

which, in view of (1.13), yields the combinatorial identity

$$(1.14) \qquad \sum_{j=1}^{m^*} \binom{M}{j}\binom{T-M}{t-j} = \binom{T}{t}$$

The identity (1.14) is known as *Euler's formula*.

Example 1.22   A lot of 15 products contains 5 defective ones. Calling a product type I if it is defective, evaluate p_m for each possible value of m, if three products are selected at random according to our model.

By assumption, Theorem 1.7 applies with $T = 15$, $M = 5$, and $t = 3$, which yields $m^* = 3$. We now compute p_0, p_1, p_2, and p_3 by (1.13). Each has the denominator

$$\binom{15}{3} = \frac{15 \times 14 \times 13}{3!} = 5 \times 7 \times 13$$

The numerators are respective products of the binomial coefficients

$$\binom{5}{0} = 1, \quad \binom{5}{1} = 5, \quad \binom{5}{2} = 10, \quad \binom{5}{3} = 10$$

and

$$\binom{10}{3} = 120, \quad \binom{10}{2} = 45, \quad \binom{10}{1} = 10, \quad \binom{10}{0} = 1$$

Hence

$$p_0 = \frac{120}{5 \times 7 \times 13} = \frac{24}{91}, \qquad p_1 = \frac{5 \times 45}{5 \times 7 \times 13} = \frac{45}{91}$$

$$p_2 = \frac{10 \times 10}{5 \times 7 \times 13} = \frac{20}{91}, \qquad p_3 = \frac{10}{5 \times 7 \times 13} = \frac{2}{91}$$

As a check, we find that $p_0 + p_1 + p_2 + p_3 = 1$.

Note the remarkable fact that p_1 is far the largest among the four values above. It is very pleasing, because its meaning is that in the most likely case, of three selected items one will be defective, which is the same ratio for defectives (1/3) as in the entire lot (5/15). In a more mathematical language, what we have is this:

In the most likely case, the relative frequency 1/3 of defectives (in this particular example) equals the probability 5/15 of selecting a defective from the original lot.   ▲

Although an exact equation between relative frequencies and probabilities is not valid in general, the next theorem does show that in the most likely case of our selection model, the relative frequency is indeed "very close" to the (abstract concept of) probability.

Theorem 1.8   For the positive integers T, M, and t, let

$$s = \frac{(t + 1)(M + 1)}{T + 2}$$

and let s^* be the unique integer satisfying $s^* \leq s < s^* + 1$.
(i)   If $s^* = s$ (i.e., if s itself is an integer), then the hypergeometric probabilities of (1.13) satisfy the inequalities

$$p_0 < p_1 < \cdots < p_{s^*-1} = p_{s^*} \qquad \text{and} \qquad p_{s^*} > p_{s^*+1} > \cdots > p_{m^*}$$

(ii) If $s^* \neq s$ (i.e., if s is not an integer), then

$$p_0 < p_1 < \cdots < p_{s^*} \qquad \text{and} \qquad p_{s^*} > p_{s^*+1} > \cdots > p_{m^*}$$

where $m^* = \min(M, t)$.

    In either case,

(1.15)
$$\left| \frac{s^*}{t} - \frac{M}{T} \right| < \frac{1}{t}$$

The proof of this theorem is given in the next section.

Note that parts (i) and (ii) of the statement above imply that s^* is a value that is most likely to occur. On the other hand, the inequality (1.15) makes the claim precise that in the most likely outcome of our selection model, the relative frequency of type I items is close to the probability M/T of selecting a type I item from the original lot. This provides a general practical method of determining the unknown value (M or T) in the models of Examples 1.19 to 1.21. This method is illustrated in Examples 1.23 and 1.24. Before these, however, we introduce a concept.

Definition 1.4   If the value of an unknown number (which is called a *parameter*) in a selection model is determined under the assumption that the result of the selection is a most likely one, we say that we have applied the *maximum likelihood principle*.

Example 1.23   In a mass production process, $t = 1000$ items are inspected at random and $m = 53$ are found to be defective. What is the true proportion of defective items in the entire lot under the maximum likelihood principle?

Since the relative frequency m/t of the defective items in the sample is $53/1000 = 0.053$, the true proportion M/T is between 5.2 and 5.4% if the maximum likelihood principle is adopted; that is, we can then appeal to (1.15).   ▲

Example 1.24   In the model of Example 1.21, $M = 50$ animals are tagged, $t = 200$ are recaptured, and $m = 8$ tagged animals are found. What is the population size under the maximum likelihood principle?

As a first approximation, we again turn to the inequalities in (1.15). We get

$$\frac{8}{200} - \frac{1}{200} < \frac{50}{T} < \frac{8}{200} + \frac{1}{200}$$

and thus $1111 < T \leqslant 1428$. However, if we utilize the full extent of Theorem 1.8, a closer range of T is obtained. Namely, with the notations of Theorem 1.8, $s = 201 \times 51/(T + 2)$ and, under the maximum likelihood principle; $s^* = 8$. Hence $8 \leqslant 201 \times 51/(T + 2) < 9$; that is, $1137 < T \leqslant 1279$. ▲

1.7 PROOF OF THEOREM 1.8

We first determine those values of i for which $p_i < p_{i+1}$. For this purpose, we evaluate the ratio p_{i+1}/p_i. We have, upon simplifying by the common denominator $\binom{T}{i}$ [see (1.13)],

$$\frac{p_{i+1}}{p_i} = \frac{\binom{M}{i+1}\binom{T-M}{t-i-1}}{\binom{M}{i}\binom{T-M}{t-i}}$$

$$= \frac{i!\,(M-i)!\,(t-i)!\,(T-M-t+i)!}{(i+1)!\,(M-i-1)!\,(t-i-1)!\,(T-M-t+i+1)!}$$

$$= \frac{(M-i)\,(t-i)}{(i+1)\,(T-M-t+i+1)}$$

Hence $p_i < p_{i+1}$, that is, $p_{i+1}/p_i > 1$ if, and only if,

$$(i+1)\,(T-M-t+i+1) < (M-i)\,(t-i)$$

Since this last inequality, after simple algebraic manipulations, becomes

$$i < \frac{Mt + M + t - T - 1}{T+2} = \frac{(t+1)\,(M+1)}{T+2} - 1 = s - 1$$

we have proved that $p_i < p_{i+1}$ for all $i < s - 1$. On the other hand, if we reverse the inequalities in the arguments above, we also obtain that $p_i > p_{i+1}$ for all $i > s - 1$ (with the obvious limitation of $i < m^*$).

In summary, we have proved so far that if s is an integer, then

$$p_0 < p_1 < p_2 < \cdots < p_{s-1} \qquad \text{and} \qquad p_s > p_{s+1} > \cdots > p_m{}^*$$

while if s is not an integer, then with the unique integer s^* satisfying $s^* < s < s^* + 1$,

$$p_0 < p_1 < p_2 < \cdots < p_{s^*} \qquad \text{and} \qquad p_{s^*} > p_{s^*+1} > \cdots > p_{m^*}$$

because if s is not an integer, the inequalities $i < s - 1$ and $i > s - 1$ are equivalent to $i \leqslant s^* - 1$ and $i \geqslant s^*$, respectively. Consequently, to prove parts (i) and (ii), it remains to show that $p_{s-1} = p_s$ whenever s is an integer. For this we go back to the formula for p_{i+1}/p_i, and we substitute $i = s - 1$, obtaining

$$\frac{p_{\iota s}}{p'_{s-1}} = \frac{(M + 1 - s)(t + 1 + s)}{s(T - M - t + s)}$$

Now substituting $s = (t + 1)(M + 1)/(T + 2)$ on the right-hand side, a lengthy but routine calculation leads to the equation $p_s/p_{s-1} = 1$, that is, $p_s = p_{s-1}$, as claimed.

To estimate the difference between s^*/t and M/T, we start with inequalities defining s^*:

$$s^* \leq \frac{(t + 1)(M + 1)}{T + 2} < s^* + 1$$

Upon dividing these inequalities by t and subtracting M/T from all terms, we get

$$\frac{s^*}{t} - \frac{M}{T} \leq \frac{(t + 1)(M + 1)}{t(T + 2)} - \frac{M}{T} < \frac{s^*}{t} - \frac{M}{T} + \frac{1}{t}$$

or, after rearranging the appropriate terms,

(1.16) $$\frac{(t + 1)(M + 1)}{t(T + 2)} - \frac{M}{T} - \frac{1}{t} < \frac{s^*}{t} - \frac{M}{T} \leq \frac{(t + 1)(M + 1)}{t(T + 2)} - \frac{M}{T}$$

We first estimate the extreme right-hand side. We have

$$\frac{(t + 1)(M + 1)}{t(T + 2)} - \frac{M}{T} = \frac{tT + MT + T - 2Mt}{tT(T + 2)}$$

$$= \frac{t(T - M) + M(T - t) + T}{tT(T + 2)}$$

$$\leq \frac{tT + T(T - t) + T}{t(T^2 + 2T)} = \frac{T^2 + T}{t(T^2 + 2T)} < \frac{1}{t}$$

We have thus proved that the extreme right-hand side of (1.16) is a positive number (see the third form in the set of equations above) which is smaller than $1/t$. Now, the extreme left-hand side of (1.16) is smaller by $1/t$ than its extreme right-hand side. Therefore, the foregoing estimate of the right-hand side implies that the absolute value of the left-hand side is also smaller than $1/t$, which completes the proof.   ▲

1.8   SELECTION WITH REPLACEMENT FROM A LOT CONTAINING TWO TYPES OF ITEMS

We now modify the selection procedure of Section 1.6 in that the items previously selected are placed back into the lot before the next selection.

That is, we again assume that we have T items, out of which M are type I and $T - M$ are type II. We select at random t items from the lot with the following procedure (which is called *selection with replacement*). One item is selected at a time and the selected item is placed back into the lot before the next selection. The outcomes (when t items have been selected) are assumed to be equally likely. Denoting by X the number of type I items among those selected, our aim is to establish the formula

$$(1.17) \qquad P(X = m) = \binom{t}{m} \left(\frac{M}{T}\right)^m \left(1 - \frac{M}{T}\right)^{t-m}$$

where $m = 0, 1, 2, \ldots, t$. The terms on the right-hand side are called the *binomial probabilities* (with parameters t and $p = M/T$).

For proving (1.17), note that, by assumption, the present selection model can be described by the classical probability space, in which a basic outcome is a t-tuple $(u_1, u_2, \ldots, u_t)$, where each u_i is any one of the T items. Therefore, the number of basic outcomes is $N = T^t$ (each u_i can take T different values). Now, $(u_1, u_2, \ldots, u_t)$ is favorable to the event $\{X = m\}$ if exactly m of the t components $u_1, u_2, \ldots, u_t$ correspond to the M type I items. We count these favorable cases in two steps. First, we select m positions (out of t) for the type I items. This can be done in $\binom{t}{m}$ ways [see (1.10) and (1.11)]. Next, when these m positions are fixed, we count the number of ways of choosing the u_j to be favorable to $\{X = m\}$. Because previously selected items are placed back before the next selection, u_j can be chosen in M ways whenever j is one of the m positions fixed for the type I items, and in $T - M$ ways for the remaining $t - m$ values of j (since these are to correspond to type II items). Consequently, this second step in our counting results in $M^m(T - M)^{t-m}$ favorable cases, and thus the number of all favorable cases (the combination of the first and second steps) is $k = \binom{t}{m} M^m(T - M)^{t-m}$. By Theorem 1.6, $P(X = m) = k/N$, which is exactly (1.17) [rewrite $N = T^t = T^m T^{t-m}$ in k/N to get the form of (1.17)].

Example 1.25   Of the 38 numbers on a roulette wheel (numbers 1 through 36 and the special numbers 0 and 00), 18 are red. Find the probability that in five spins of the wheel a red number wins exactly twice.

Playing roulette is a typical example of selection with replacement. In this particular case, $T = 38$, $M = 18$, and $t = 5$. The formula (1.17) thus gives

$$P(X = 2) = \binom{5}{2} \left(\frac{18}{38}\right)^2 \left(\frac{20}{38}\right)^3 = 10 \times 0.224 \times 0.146 = 0.327 \qquad \blacktriangle$$

Example 1.26   Toss a fair coin eight times. Find the probability that the number X of times when head turns up is either 3, or 4, or 5.

It is again a problem of selection with replacement, where $T = 2$ (head or tail), $M = 1$ (head), and $t = 8$. Our interest is

$$P(X = 3 \text{ or } 4 \text{ or } 5) = P(X = 3) + P(X = 4) + P(X = 5)$$

the equation being valid on account of axiom $(A3)$ of probability. Now, by (1.17),

$$P(X = 3) = \binom{8}{3} \left(\frac{1}{2}\right)^3 \left(\frac{1}{2}\right)^5 = \frac{56}{2^8}, \qquad P(X = 4) = \binom{8}{4} \left(\frac{1}{2}\right)^4 \left(\frac{1}{2}\right)^4 = \frac{70}{2^8}$$

and

$$P(X = 5) = \binom{8}{5} \left(\frac{1}{2}\right)^5 \left(\frac{1}{2}\right)^3 = \frac{56}{2^8}$$

and thus the answer is $182/2^8 = 0.711$.   ▲

With earlier terminology, X is the frequency of heads in eight repetitions of tossing a fair coin, and thus $X/8$ is its relative frequency. From the fact that $P(\text{head in one toss}) = 1/2$ we cannot expect that $X/8 = 1/2$ (i.e., $X = 4$), but the result of the last example shows that with high probability (0.711), X is "close to" 4. The strong relation between X/t and $1/2$ can better be recognized if we increase the value of t. In Chapter 2 we shall be able to compute that, when $t = 1600$, say, $0.475 < X/t < 0.525$ with probability 0.95. [We can easily give a formula for this probability by an appeal to (1.17), but the binomial coefficients $\binom{1600}{m}$ appearing in such a formula with m around 800 require lengthy computations even on a computer.]

In the following example, the selection models with and without replacement are compared.

Example 1.27   Out of the integers 1 through 100, on one occasion, five distinct numbers are picked at random, while on another, five numbers are picked at random, but this time repetitions are permitted. Let us evaluate in both cases the probability p_2 that exactly two even numbers are picked.

In both cases the numbers 1 through 100 are split into two types: even or odd. Thus, when insisting on getting distinct numbers, the model of selection without replacement applies, whereas the second case, when repetitions are permitted, leads to the model of selection with replacement. Hence, in the first case, by Theorem 1.7,

$$p_2 = \frac{\binom{50}{2}\binom{50}{3}}{\binom{100}{5}} = 0.319$$

and for the second case (1.17) yields

$$p_2 = \binom{5}{2}\left(\frac{50}{100}\right)^2\left(1 - \frac{50}{100}\right)^3 = \frac{10}{32} = 0.313 \qquad \blacktriangle$$

The two values above are remarkably close, which is not an accident. The hypergeometric probabilities of Theorem 1.7 and the binomial probabilities of (1.17) are generally close to each other whenever t/T is "small." In fact, the approximation by the binomial probabilities (1.17) to the hypergeometric probabilities (1.13) is usually adequate if $t < 0.1\ T$. We do not justify this claim here; a good collection of results on this approximation can be found in Johnson and Kotz, *Discrete Distributions* (Wiley, New York, 1969, pp. 148-151).

We shall return to the binomial probabilities in Chapter 2.

1.9   FURTHER MODELS WITH CLASSICAL PROBABILITY SPACES

We first extend the selection model of Section 1.6 to the case when there are more than two types of items in the lot. The necessity of such extension arises if we want to understand our risk in simple card games in casinos, but the beauty of mathematical abstraction is reflected in the fact that the same model enables us to answer several other practical questions as well. The following two examples, which we solve by a single abstract argument, make the preceding claim specific.

Example 1.28   A deck of cards consists of four suits, and each suit consists of 13 denominations. Let us call six denominations "low-value cards," four "high-value cards," and the remaining three "neutral." From a well-shuffled deck of cards, six are dealt out. Let us find the probability that these six cards contain three low values, one neutral, and two high values. In addition, let us determine the probability that the six dealt-out cards contain two low values from the suit of spades, and three high values from any suit.

Before solving this problem, we set up another one.

Example 1.29   From the point of view of taxation, the population is split into three categories: low-income, middle-income, and high-income families. If from a population of 80 million families, 2000 are selected at random (without replacement) for a thorough tax audit, what is the probability that the three categories will be represented in the same proportion in the sample as in the entire population? (To be specific, assume that 20% of the population are low-income families and that 65% belong to the middle-income group.)

Notice that the questions of Examples 1.28 and 1.29 can be reformulated by means of the following single abstract model. Let an urn contain T balls, out of which n_1 are red, n_2 are white, and n_3 are blue $(n_1 + n_2 + n_3 = T)$. Select t balls from the urn at random and without replacement. Our interest is to find the probability that we selected t_1 red, t_2 white, and t_3 blue balls $(t_1 + t_2 + t_3 = t)$.

The solution should be routine for us by now. We face a classical model, and thus we have to count the number of all possible outcomes and of those which are favorable to the event we specified. Just as in Theorem 1.7, the number of all possible outcomes is the binomial coefficient $\binom{T}{t}$. For a selection to be favorable, we have to choose t_1 of the n_1 red, t_2 of the n_2 white, and t_3 of the n_3 blue balls. The number of such possible choices is the product of the binomial coefficients $\binom{n_i}{t_i}$, $1 \leq i \leq 3$, and thus the desired probability equals

$$(1.18) \qquad \frac{\binom{n_1}{t_1}\binom{n_2}{t_2}\binom{n_3}{t_3}}{\binom{T}{t}}$$

Now, the solutions to the problems of Examples 1.28 and 1.29 are straightforward from (1.18).

As a matter of fact, if we replace the word "ball" by "card," "red" by "low-value card," "white" by "high-value card," and "blue" by "neutral," then the substitution $T = 52$, $t = 6$, $n_1 = 24$, $n_2 = 16$, $n_3 = 12$, $t_1 = 3$, $t_2 = 2$, and $t_3 = 1$ in (1.18) gives the answer to the first question of Example 1.28:

$$\frac{\binom{24}{3}\binom{16}{2}\binom{12}{1}}{\binom{52}{6}}$$

On the other hand, if we change our translation of "red" to "low-valued spades" but keep "white" to mean "high-value card" (thus "blue" is now any card not yet classified, i.e., either neutral or low-valued but not spades), then the answer to the second question of Example 1.28 is (1.18), again with $T = 52$, $t = 6$, $n_1 = 6$, $n_2 = 16$, $n_3 = 30$ (recall that $n_1 + n_2 + n_3 = T$), $t_1 = 2$, $t_2 = 3$, and $t_3 = 1$; the actual value is

$$\frac{\binom{6}{2}\binom{16}{3}\binom{30}{1}}{\binom{52}{6}}$$

Finally, if "family" is "ball" in the urn model above, and the three colors red, white, and blue are identified with the low-, middle-, and high-income families, respectively, then the answer to Example 1.29 is once more (1.18), where $T = 80 \times 10^6$, $t = 2000$, $n_1 = 16 \times 10^6$, $n_2 = 52 \times 10^6$, $n_3 = 12 \times 10^6$, $t_1 = 400$, $t_2 = 1300$, and $t_3 = 300$. This answer, however, even after the substitution, is symbolic only, because the resulting binomial coefficients are almost impossible to compute. As remarked earlier, this will be overcome by approximation methods in Chapter 2. [Note: In this solution, we took t_1 and t_2 as exactly 20% and 65%, respectively, of $t = 2000$, but evidently no one would argue that the sample had the required proportion if $t_1 = 386$ and $t_2 = 1322$, say. This means that in order to give a practical answer to Example 1.29, one has to utilize (1.18) with several values of t_1, t_2, and t_3, all of which are reasonably close to the single values above. The sum of all these probabilities would be a practical answer.]

Formula (1.18) can easily be extended to an urn model in which there are balls of k different colors; see Exercise 35. Urn models are convenient tools for solving practical problems. The following problem originates from physics.

Assume that m identical balls are distributed into n urns. Every distribution of the balls describes a possible phase of a physical system, and every such distribution of the balls is equally likely (this assumption is known in physics as the Bose-Einstein statistic). What is the probability that the system is in a given phase?

From the assumption of equally likely outcomes, the answer is $1/N$, where N is the number of ways of distributing m identical balls in n (distinct) urns. Let us label the urns by 1 through n, and let x_j be the number of balls in urn j. Then N is the number of solutions of the equation

(1.19) $x_1 + x_2 + \cdots + x_n = m$ $x_j \geq 0$ integer

Introduce $y_j = x_j + 1$. Then

(1.20) $y_1 + y_2 + \cdots + y_n = m + n$ $y_j \geq 1$ integer is equivalent to (1.19). For solving (1.20), imagine that the positive integers $y_1, y_2, \ldots, y_n$ are placed successively on the real line, starting at zero. We then identified n points: $y_1, y_1 + y_2, \ldots, y_1 + y_2 + \cdots + y_n$. The last of these points is always $m + n$, from (1.20), and the other $n - 1$ points can be any of the integers $1, 2, \ldots, m + n - 1$. Consequently, the number N of solutions of (1.20), and thus of (1.19) as well, is the number of ways of choosing $n - 1$ out of $n + m - 1$; we know this to be the binomial coefficient $\binom{n+m-1}{n-1}$. Therefore, under the assumption of the Bose-Einstein statistic, every phase of the system has probability $1/\binom{n+m-1}{n-1}$ to occur.

It is important to emphasize that the assumption of equally likely outcomes in a physical model such as the one described above is just an assumption, and there is no way of checking its exact validity. Rather, several conclusions are drawn from such an assumption, and if observations are in agreement with those conclusions, the assumption is accepted. It turned out that the behavior of photons is well described by the Bose-Einstein statistic, but the assumption of this statistic has to be modified in connection with electrons and protons. The latter seem to follow that classical probability model in which the Bose-Einstein assumption is modified by including the additional assumption that every urn can contain at most one ball. This model is known in physics as the Fermi-Dirac statistic, in which the system is in a given phase with probability $1/\binom{n}{m}$ (why?).

1.10   FINITE AND DENUMERABLE SAMPLE SPACES

Even if the outcomes in a finite sample space are not equally likely, there is a simple way of computing probabilities. For the method that we are to describe in the present section, however, it makes no difference whether the sample space is finite or denumerably infinite. The method is as follows.

First, identify the elements of the sample space Ω: $\Omega = \{a_1, a_2, \ldots\}$. Then compute $p_k = P(\{a_k\})$, $k = 1, 2, \ldots$ (i.e., for each possible outcome a_k, we have to be able to compute the probability that $\{a_k\}$ occurs). As a check, we have to have

$$(1.21) \qquad p_k \geqslant 0, \qquad p_1 + p_2 + \cdots + p_n + \cdots = 1$$

Next, for a given event A, which has to be a subset of Ω, identify those members a_{i_k} of Ω which are favorable to A: that is, the precise set

$$(1.22) \qquad A = \{a_{i_1}, a_{i_2}, \ldots\}$$

as a subset of Ω should be identified. Then

$$(1.23) \qquad P(A) = p_{i_1} + p_{i_2} + p_{i_3} + \cdots$$

The justification of this computation method is immediate from the axioms of probability. By assumption,

$$\Omega = \{a_1\} \cup \{a_2\} \cup \cdots$$

where the events on the right-hand side are mutually exclusive. Hence, by axioms $(A2)$ and $(A3)$, equation (1.21) follows, while the inequality $p_k \geq 0$ is just axiom $(A1)$. Finally, since (1.22) is equivalent to $A = \{a_{i_1}\} \cup \{a_{i_2}\} \cup \cdots$, one more appeal to axiom $(A3)$ yields (1.23).

Example 1.30   In an experiment, there are only three possible outcomes: a_1, a_2, and a_3. The event $\{a_1\}$ is twice as likely as $\{a_2\}$, and $\{a_2\}$ is three times as likely as $\{a_3\}$. Let us determine $p_j = P(\{a_j\})$, $1 \leq j \leq 3$.

By assumption, $p_1 = 2p_2$ and $p_2 = 3p_3$. Substitution into (1.21) thus gives $6p_3 + 3p_3 + p_3 = 1$, from which $p_3 = 0.1$. Hence $p_1 = 0.6$ and $p_2 = 0.3$.   ▲

Before going to the next example, let us record the finite and infinite summation formula for the *geometric series*. These are

$$(1.24) \qquad 1 + x + x^2 + \cdots + x^n = \frac{1 - x^{n+1}}{1 - x} \qquad x \neq 1$$

and

$$(1.25) \qquad \sum_{j=0}^{+\infty} x^j = \frac{1}{1 - x} \qquad |x| < 1$$

The proof of these formulas is very simple. Let s_n represent the left-hand side of (1.24). Because in $s_n - xs_n$ most terms cancel out, we get the simple identity

$$s_n(1 - x) = s_n - xs_n = 1 - x^{n+1}$$

which gives (1.24). Now, since

$$\sum_{j=0}^{+\infty} x^j = \lim_{n = +\infty} \sum_{j=0}^{n} x^j$$

(1.24) implies (1.25).

Example 1.31   Assume that in an experiment the possible outcomes are the positive integers, and $p_k = P(\{k\}) = c3^{-k}$, where $c > 0$ is the same value for all $k \geq 1$. Let us determine the probability that the outcome in this experiment is larger than 3.

We utilize (1.21) for the evaluation of c. By substitution,

$$1 = c \sum_{k=1}^{+\infty} 3^{-k} = \frac{c}{3} \sum_{k=1}^{+\infty} \frac{1}{3^{k-1}} = \frac{c}{3} \frac{1}{1 - 1/3} = \frac{c}{2}$$

where, in the next-to-last equation, we applied (1.25) with $x = 1/3$. We thus have $c = 2$. Now, if we put $A = \{\text{outcome} > 3\}$, then $A = \{4, 5, 6, \ldots\}$, and thus by (1.23) [and once again by (1.25)],

$$P(A) = \sum_{k=4}^{+\infty} \frac{2}{3^k} = \frac{2}{3^4} \sum_{t=0}^{+\infty} \frac{1}{3^t} = \frac{2}{3^4} \frac{1}{1 - 1/3} = \frac{1}{27} \qquad \blacktriangle$$

The situation of Example 1.31 arises in waiting-time problems. Since we shall discuss waiting times in their generality in Chapter 3, here we look at a simple case only.

Let an urn contain one white and two red balls. Draw balls from the urn at random with replacement and stop as soon as a red ball is drawn. Let Ω represent the number of balls required to be drawn by this experiment (in other words, we wait for a red ball to be drawn). Then every positive integer is a member of Ω. Now, the outcome $\{k\}$ means that in the first $k - 1$ choices, we repeatedly drew white balls, and the kth choice resulted in a red one. Therefore, $P(\{k\})$ can be computed by the classical probability space of drawing exactly k balls with replacement; in this there are 3^k possible outcomes, and the number of favorable cases is 2; that is, in the first $k - 1$ places we have a single possibility (one white at each time), and the kth choice can be either of the two red balls. Hence $P(\{k\}) = 2/3^k$, which is exactly the value of p_k in Example 1.31.

1.11 GEOMETRIC PROBABILITIES

The term "geometric probability" comes from the assumption that we deal with random choices of points from a geometric region, and we measure probability in the same manner in which the corresponding region is measured in geometry. Thus, if we pick a point on a segment of the real line, probability will be proportional to length, whereas when a point is picked in a region of the plane, probability is proportional to area; and so on. Note that such choices of probability are in complete agreement with the axioms of probability, because geometric measures are always nonnegative, and the union of disjoint sets is measured as the sum of the measurements of the individual terms. Axiom $(A2)$ is the only one that is not automatically satisfied, but this can be taken care of by the coefficient of proportion between probability and the appropriate geometric measure.

Example 1.32 Pick a point x at random on the interval $(0, 3)$. Let us determine the probability that $1 \leqslant x \leqslant 2.5$.

Since x is picked in an interval, by assumption,

$$P(1 \leq x \leq 2.5) = c(2.5 - 1) = 1.5c$$

with some c (since probability is proportional to length). Next, we compute c by utilizing axiom $(A2)$. Since $\Omega = (0, 3)$,

$$1 = P(\Omega) = c(3 - 0) = 3c$$

i.e., $c = 1/3$. Hence $P(1 \leq x \leq 2.5) = 1.5/3 = 1/2$.   ▲

One reason behind this answer of 1/2 is that one-half of $\Omega = (0, 3)$ is favorable to $1 \leq x \leq 2.5$. This explanation reflects well the meaning of geometric probabilities: The actual location of the interval (1, 2.5) is irrelevant; only its length (1.5) is significant for computing the probability in question. In other words, if we view the classical model of equally likely outcomes in a finite sample space as a model in which no outcome is preferred to the others, we can say that the geometric probabilities are the appropriate extensions of (finite) classical spaces to (finite segments of) the line, plane, and so on.

What is the value of $P(x = 1)$ in Example 1.32? Since $\{x = 1\} = \{1 \leq x \leq 1\}$, whose length is zero, $P(x = 1) = c \times 0 = 0$. We get the same answer (zero) for $P(x = a)$, whatever the value of $a \in \Omega$; that is, every individual outcome of Ω is of zero probability. Consequently, if we modify a set A by a finite or denumerably infinite number of points (deleting them from A or adding them to A), the probability of whether x falls into A or into the modified set is the same [in view of axiom $(A3)$]. Such a phenomenon is possible only because Ω is infinite and *not* denumerable.

Example 1.33   Pick a point (x, y) in the unit square $\Omega = \{(x, y): 0 \leq x \leq 1, 0 \leq y \leq 1\}$. Let us find the probability that $x \leq y \leq 1 - x$.

Here probability is proportional to area. But because in this particular case, both the area and the probability of Ω equal 1, probability is in fact equal to "area." Now, $y = x$ and $y = 1 - x$ are the two diagonals of Ω; hence the set $\{(x, y): x \leq y \leq 1 - x\}$ is that triangle in Ω which lies below $y = 1 - x$ and above $y = x$ (the shaded region of Figure 1.4). Its area is 1/4; that is, $P(x \leq y \leq 1 - x) = 1/4$.   ▲

When more than one points are picked from a region, the points are transformed into an appropriate higher-dimensional space for calculating probabilities.

Example 1.34   Pick two points x and y on the unit interval (0, 1). Let us find the probability that $|x - y| < 1/2$.

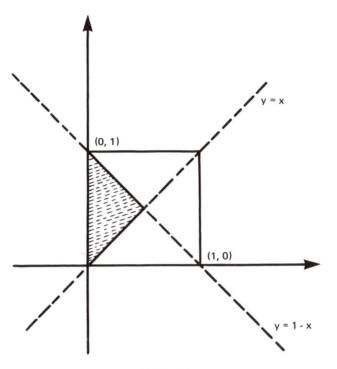

Figure 1.4

Picking two points x and y on $(0, 1)$ is viewed as picking one point (x, y) in the square $\Omega = \{(x, y): 0 < x < 1, \ 0 < y < 1\}$. Now, the set $\{(x, y): |x - y| < 1/2\} = \{(x, y): -1/2 < x - y < 1/2\}$ as a subset of Ω is the set of those points of the square Ω which lie both below the line $y = x + 1/2$ and above the line $y = x - 1/2$ (the shaded region of Figure 1.5). Since the two unshaded triangles of Ω have area 1/8 each, the shaded area is $1 - 2/8 = 3/4$, which also equals the desired probability (because the area of Ω is 1). ▲

In the geometric approach to calculating probabilities, questions are sometimes ambiguous. This is the case in the following example, which is known as the *Bertrand paradox*.

Example 1.35 Let L be the length of the side of the equilateral triangle inscribed in circle C. Pick a chord A of C at random. What is the probability that the length a of A is larger than L?

Because the "experiment" of picking a chord at random is not uniquely determined, we get the value of the desired probability according to our interpretation of the mentioned experiment. For example, one can pick A

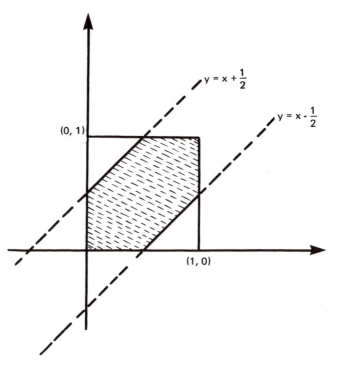

Figure 1.5

by picking its middle point as an arbitrary point of C. In this case, probability is proportional to area, and simple calculation leads to $P(a > L) = 1/4$. On the other hand, we can refer to the symmetry of the circle and say that we can fix a radius and we can pick the middle point of A on this radius. With this approach, probability is proportional to length, and we get $P(a > L) = 1/2$. A third possibility is to choose the endpoints of A on the circumference of C. Because of symmetry, this means that one endpoint of A can be fixed and the other one is chosen at random on the circumference of C. Probability is again proportional to length, and one gets $P(a > L) = 1/3$. The details of these calculations are left to the reader (it is advised to exploit the symmetry of the equilateral triangle and to place it in an appropriate position in the circle for simplifying calculations).   ▲

1.12 EXERCISES

1. Give a set Ω as the sample space for each of the following experiments:
   (i)   roll three (distinguishable) dice;
   (ii) toss a coin and roll a die;

(iii) select three (distinct) numbers from the integers 1 through 100.

2. The set $\Omega = (0, 1)$ is chosen as the sample space for tossing a coin. Give a dictionary identifying heads and tails. Using your dictionary, give a subset of Ω which is not an event.

3. Explain why the set $S = \{1, 2, 3, ..., 35\}$ cannot be chosen as the sample space for rolling two distinguishable dice.

4. Can the set $\Omega = \{1, 2, 3, ..., 35\}$ be the sample space for rolling two identical dice? At least how many elements should a sample space contain for this experiment?

5. In each of the following experiments choose a sample space Ω and identify the event A (specified below) as a subset of Ω:
   (i)   roll two (distinguishable) dice:
         $A = \{\text{the sum of the faces is less than 8}\}$
   (ii) pick a point in a square (whose edges are the coordinate axes):
         $A = \{\text{the point chosen is closer to the } x \text{ axis than to the } y \text{ axis}\}$

6. For rolling two distinguishable dice, the set $\Omega = \{(i, j): 1 \leqslant i \leqslant 6, 1 \leqslant j \leqslant 6\}$ is chosen as the sample space. Express by words the events $A = \{(1, 1), (3, 3), (5, 5)\}$ and $B = \{(1, 2), (2, 1), (2, 2)\}$.

7. A number x is chosen from the integers 1, 2, ..., 15. Which special events are represented by $\{x < 18\}$ and $\{x > 16\}$?

8. The experiment of choosing a positive number is represented by $\Omega = (0, +\infty)$. For the events $A = (2, 8.5)$ and $B = (5.8, 12)$, find (i) $A^c \cap B$; (ii) $A \cup B$; and (iii) $A \cap (A^c \cap B)$.

9. Show that, for arbitrary events A, B, and C,

$$A \cup (B \cap C) = (A \cup B) \cap (A \cup C)$$

10. Express with the symbols of union, intersection, and complement the event that exactly two of the events A, B, and C occur.

11. Verify, either by a Venn diagram or by utilizing the basic rules, that for arbitrary events A and B,

$$(A \cup B^c) \cap (A \cap B)^c = B^c$$

12. Every weekday night three digits, each from 0 to 9, are drawn live on a TV channel, yielding the Daily Number. In a month, the actual Daily Numbers were: 826, 663, 232, 662, 053, 551, 223, 545, 786, 976, 444, 626, 354, 752, 118, 653, 023, 007, 569, 927, 690, 109, 312, 247, and 301. Let A be the event that the Daily Number contains the digit 6, B that it is even, and C that among the three digits of the Daily Number two consecutive integers appear. Count the frequency of A, B, C, and $A \cap B^c$.

13. $P(A) = 0.8$, $P(B) = 0.2$, and $P(A \cap B) = 0.1$. Evaluate (i) $P(A^c)$; (ii) $P(A \cap B^c)$; and (iii) $P(A^c \cup B)$.

14. There is a contradiction among the probabilities $P(A) = 0.8$, $P(B) = 0.2$, and $P(A \cup B^c) = 0.7$. Why?

15. Show that if $P(A) = 0.3$ and $P(B) = 0.4$, then $0.4 \leqslant P(A \cup B) \leqslant 0.7$. Can the two bounds be achieved?

16. In a group of 120 students, 76 are enrolled in a course in probability theory, 52 in a course in algebra, and 28 in both. If a student is selected at random from this group, what is the probability that the student is enrolled in neither course?

17. A restaurant offers 8 different salads, 12 different main dishes, and 6 desserts. In how many different ways can a customer choose a three-course meal (salad, main dish, and dessert)?

18. In how many ways can a chairman, vice chairman, a secretary, and a treasurer be chosen from among the 10 members of the governing body of a club?

19. The numbers 1, 2, 3, 4, and 5 are put down in random order, yielding a five-digit number. What is the probability that (i) the first digit of this number is 1; (ii) the last digit of this number is 5; (iii) the first digit is 1 and the last digit is 5; and (iv) the first digit is 1 and the last digit is not 5?

20. On a key holder there are five keys of the same shape. If only one of the keys opens the main entrance, what is the probability that a person arriving home in the dark will need exactly three attempts to open the door? What about one attempt, or five attempts?

21. If 15 people, including A and B, are seated at random (i) in a row (ii) at a round table, what is the probability that A and B are seated adjacent to each other?

22. Assume that two boys and two girls are seated at random in a row. Find the probability that the boys and the girls sit in alternate seats.

23. Compute the binomial coefficients $\binom{18}{3}$ and $\binom{18}{16}$.

24. Show that, for integers $0 \leqslant k \leqslant n$,

$$\binom{n}{k} + \binom{n}{k+1} = \binom{n+1}{k+1}$$

25. Show that, for $1 \leqslant k < n$,

$$k\binom{n}{k} = n\binom{n-1}{k-1}$$

26. Show that for an arbitrary nonnegative integer b,

$$\sum_{t=0}^{b} (-1)^t \binom{n}{t} = (-1)^b \binom{n-1}{b}$$

27. In how many ways can a committee of five be selected from the 100 U.S. senators?

28. In a class of eight male and seven female students, a committee of five is selected at random. What is the probability that the committee contains three male students?

29. A deck of 52 cards has 13 spades. If from a well-shuffled deck, five cards are dealt out, what is the probability that these five contain at most two spades?

30. An urn contains 50 red and 50 white balls. Four balls are selected at random

(i) with replacement and (ii) without replacement. Find the probability in each case that exactly two red balls are obtained.

31. From an urn containing M red and $T - M$ white balls, t balls are selected at random, one by one and without replacement. Let A_j, $1 \leqslant j \leqslant t$, be the event that the jth selection results in a red ball. Show that
    (i) $P(A_j) = M/T$ for each value of j
    (ii) $P(A_k \cap A_j) = M(M - 1)/T(T - 1)$ for all $k \neq j$

32. Which is more likely: (i) rolling a regular die four times, 6 turns up at least once, or (ii) rolling two regular dice 24 times, a double 6 is obtained at least once?

33. A pair of fair dice is tossed, and x and y denote the two numbers turning up. Find the probability that (i) $x = y$; (ii) the larger of x and y is smaller than 4; and (iii) $|x - y| = 1$.

34. A deck of 52 cards consists of four suits with an equal number of cards in each suit. If from a well-shuffled deck, 13 cards are dealt out, what is the probability that these 13 cards contain 4, 4, 3, and 2, respectively, from suits 1, 2, 3, and 4?

35. Generalize Exercise 34 as follows. Let an urn contain n_i balls of color i, where i is one of k different colors. Select at random, without replacement, t balls from the urn. Give a formula for the probability that t_i balls of color i are among those selected ($t_1 + t_2 + \cdots + t_k = t$).

36. A deck of 52 cards contains four of each of 13 denominations. From a well shuffled deck 13 cards are dealt out. Find the probability that these 13 cards contain (i) one from each denomination, and (ii) all four cards of three denominations.

37. A city purchases 12 police cars and distributes them at random to four districts. What is the probability that one specific district gets just one car?

38. Assume m identical balls are distributed into n urns in such a way that every distribution of the balls has the same probability (the Bose-Einstein statistic). Find the probability that a given urn contains at most d balls.

39. Continuing Exercise 38, find the probability that exactly k urns remain empty.

40. What is the probability that under the assumption of the Bose-Einstein statistic, the aggregate number of balls in three specific urns is s?

41. There are three male and five female candidates for an elective office. Those of the same sex have equal probabilities of winning, but each man is twice as likely to win as any woman. Find the probability that a woman wins the election.

42. Assume an experiment is represented by $\Omega = \{1, 2, 3, 4, 5\}$. Find the probability that the result is an even number if $P(\{1\}) = 0.1$, $P(\{2\}) = 0.3$, $P(\{3\}) = 0.2$, and $P(\{4\}) = P(\{5\})$.

43. A pair of fair dice is rolled repeatedly. Let A_n be the event that the sum of the upturned faces is 6 on the nth roll and neither 6 nor 7 occurred as the sum of the two faces on the first $n - 1$ rolls. Find $P(A_n)$. Interpret the event $\bigcup_{n=1}^{+\infty} A_n$, and evaluate its probability with an appeal to axiom $(A3)$.

44. Pick two points x and y at random on the interval $(0, 3)$. Find the probability that (i) $x < y$, and (ii) $x = y$.

45. Pick two points on the unit interval $(0, 1)$, which split the interval into three

segments. What is the probability that these three segments form a triangle?

46. Three points are selected at random on the circumference of a circle. Find the probability that the points lie on a semicircle.

47. Pick a point (x, y) at random in the triangle whose vertices are the origin and the points $(4, 0)$ and $(3, 2)$. Find the probability of the events (i) $x < 2$, (ii) $x < 2y$, and (iii) (x, y) is in the rectangle with vertices $(1, 0)$, $(1, 2)$, $(2, 2)$, and $(2, 0)$.

2

Conditional Probability and Independence

2.1 THE DEFINITION OF CONDITIONAL PROBABILITY

When making a travel plan two years ahead, say, the only way of predicting the weather for the period of travel is to get information from a data bank for the past several years, then by computing relative frequencies, assign probabilities to specific weather conditions. (Although the approximate equality between relative frequencies and the corresponding probabilities is yet to be established, we assume its validity for this introduction.)

To be more specific, assume that the planned vacation is for the period of July 10 through 19. We receive complete information for this period for the past 50 years, and we find that in 46 years the weather conditions were pleasant during this period. Hence the relative frequency 46/50 = 0.92 is taken as the probability of having pleasant weather during the period July 10 through 19 in any future year, and on this basis, we can decide to take the vacation. Now, as the time approaches, we follow the weather conditions closely, and it turns out that the last week of June has been very unpleasant (hot and humid with frequent rains, say). How can this new information be combined with the experience of the past 50 years to make our prediction more reliable?

The most evident way is as follows. We go back to the data bank and select those years only in which the last week of June was comparable with our recent experience. (38 of the available 50 years, say). From these 38 years we then count the number of years in which the weather for July 10 through 19 was what we term "pleasant" (36 years, say). Therefore, the

43

new relative frequency, which takes into account our most recent experience, is $36/38 = 0.95$, which is our "new probability" utilizing our new information.

Let us rewrite the preceding discussion using mathematical symbols. Let A be the event that at a specific resort the weather is "pleasant" (which we specify for ourselves) for July 10 through 19, and let B be the event that, at this same resort the weather during the last week of June is the same as experienced most recently. Now, originally the records of $n = 50$ years were available through the data bank, but when B occurred, we discarded the records of several years and we actually kept only those in which B occurred. In other words, the number of years utilized in the computation of our "new probability" is exactly the frequency $k_B(50) = 38$ of B in 50 years. Out of these we selected those years in which not only B but also A occurred, that is, the frequency $k_{A \cap B}(50) = 36$ of $A \cap B$ in 50 years. Therefore, the relative frequency, given that B has occurred, is $k_{A \cap B}(n)/k_B(n)$, where $n = 50$. Consequently, if the relative frequency and the probability of an event are in fact close to each other as n increases, then the probability of A, given that B has occurred, denoted $P(A|B)$, should be close to $k_{A \cap B}(n)/k_B(n)$. On the other hand, if no information is available, then for every event C, $k_C(n)/n$ is approximately $P(C)$, and thus

$$(2.1) \qquad P(A \mid B) \sim \frac{k_{A \cap B}(n)}{k_B(n)} = \frac{k_{A \cap B}(n)/n}{k_B(n)/n} \sim \frac{P(A \cap B)}{P(B)}$$

assuming that $P(B) > 0$. Therefore, if we do not want to contradict (2.1) (or the argument leading to it), then the only way of defining the conditional probability $P(A \mid B)$ is as follows.

Definition 2.1   Let A and B be two events in connection with the same random experiment. Assume that $P(B) > 0$. Then the conditional probability $P(A \mid B)$ of A, given that B has occurred, is defined by the formula

$$(2.2) \qquad\qquad\qquad P(A \mid B) = \frac{P(A \cap B)}{P(B)}$$

Example 2.1   From an urn containing two red and three white balls, pick two balls at random, one by one and without replacement. Let us determine the probability that the second ball drawn is red, given that the first one was red as well.

Let $A = \{$second is red$\}$ and $B = \{$first is red$\}$. Then $P(B) = 2/5$ and $P(A \cap B) = (2 \times 1)/(5 \times 4) = 1/10$. Hence, by (2.2), $P(A \mid B) = 0.1/0.4 = 1/4$.

As a second solution, let us now argue directly. Since we know that B

has occurred, we in fact know that before the second choice, the urn contains one red and three white balls. Therefore, $P(A \mid B)$ should be 1/4, confirming our formula (2.2) to be the only reasonable one for conditional probabilities.   ▲

When the definition (2.2) of conditional probabilities is applied in the form

(2.3) $$P(A \cap B) = P(B)P(A \mid B)$$

it frequently facilitates the computation of the probability of the intersection of two events.

Example 2.2   From an urn containing two red and three white balls pick one ball at random, check its color, and place the ball, together with another ball of the same color, back into the urn. Now draw a second ball at random. Find the probability that both balls drawn are red.

Again let $A = \{$second is red$\}$ and $B = \{$first is red$\}$. Evidently, $P(B) = 2/5$. By direct computation, $P(A \mid B) = 3/6$, because, given B, the urn contains three red and three white balls before the second choice. Hence, by (2.3),

$$P(\text{both are red}) = P(A \cap B) = \frac{2}{5} \times \frac{3}{6} = \frac{1}{5} \qquad ▲$$

In both Examples 2.1 and 2.2 we treated the conditional probability $P(A \mid B)$ as a probability in a modified experiment (in which B is taken into account). In other words, we assumed implicitly that axioms (A1) to (A3) apply to $P(A \mid B)$. This assumption is correct, which is established below.

Theorem 2.1   Let a random experiment be described by the probability space $(\Omega, \mathcal{A}, P)$. Let $B \in \mathcal{A}$ with $P(B) > 0$. Then $P(A \mid B)$, defined by (2.1) for all $A \in \mathcal{A}$, satisfies axioms (A1) to (A3) of probability when the role of Ω is replaced by B, and thus every event A is transformed into a subset of B by replacing A by $A \cap B$.

Proof: For (A1), we have to show that $0 \leq P(A \mid B) \leq 1$. The lower inequality is evident; the upper inequality, on the other hand, follows from Theorem 1.3 upon observing that $A \cap B \subset B$, and thus $P(A \cap B) \leq P(B)$.

Because $B \cap B = B$, $P(B \mid B) = 1$, which establishes (A2).

Turning to axiom (A3), let $A_1, A_2, \ldots$ be mutually exclusive events. Then the events $A_1 \cap B, A_2 \cap B, \ldots$ are also mutually exclusive, and thus

$$P(\bigcup_{j=1}^{+\infty} A_j|B) = \frac{P((\bigcup_{j=1}^{+\infty} A_j) \cap B)}{P(B)} = \frac{P(\bigcup_{j=1}^{+\infty} (A_j \cap B))}{P(B)}$$

$$= \sum_{j=1}^{+\infty} \frac{P(A_j \cap B)}{P(B)} = \sum_{j=1}^{+\infty} P(A_j|B)$$

which establishes $(A3)$ for $P(A \mid B)$. The theorem is proved.   ▲

2.2 THE MULTIPLICATION RULE FOR INTERSECTIONS

The rule (2.3) for computing the probability of an intersection can be extended to an arbitrary number of terms.

Theorem 2.2 If the events A_1, A_2, and A_3 are such that $P(A_1) > 0$ and $P(A_1 \cap A_2) > 0$, then

(2.4) $P(A_1 \cap A_2 \cap A_3) = P(A_1)P(A_2 \mid A_1)P(A_3 \mid A_1 \cap A_2)$

In general, for events A_1, A_2, ..., A_n such that $P(A_1 \cap A_2 \cap \cdots \cap A_k) > 0$ for $1 \le k \le n - 1$,

(2.5)
$$P(A_1 \cap A_2 \cap \cdots \cap A_n) = P(A_1)P(A_2 \mid A_1) \cdots P(A_n \mid A_1 \cap A_2 \cap \cdots \cap A_{n-1})$$

Proof: The theorem follows easily from the definition (2.2) of conditional probabilities. Thus, since

$$P(A_2 \mid A_1) = \frac{P(A_1 \cap A_2)}{P(A_1)} \quad \text{and} \quad P(A_3 \mid A_1 \cap A_2) = \frac{P(A_1 \cap A_2 \cap A_3)}{P(A_1 \cap A_2)}$$

in the product $P(A_1)P(A_2 \mid A_1)P(A_3 \mid A_1 \cap A_2)$, every term cancels out except the last numerator, which establishes (2.4). Similarly, when we substitute (2.2) for the conditional probabilities on the right-hand side of (2.5), all terms except the last numerator will cancel out, leading to the equation (2.5).   ▲

Example 2.3 From a well-shuffled deck of cards, four cards are dealt out one by one. Find the probability that the following cards are obtained: a 2, a 3, an 8, and then another 8 (in this order).

Put $A_1 = \{$first is 2$\}$, $A_2 = \{$second is 3$\}$, $A_3 = \{$third is 8$\}$, and $A_4 = \{$fourth is 8$\}$. Then $P(A_1) = 4/52$, $P(A_2 \mid A_1) = 4/51$, $P(A_3 \mid A_1 \cap A_2) = 4/50$, and $P(A_4 \mid A_1 \cap A_2 \cap A_3) = 3/49$. Thus, by (2.5),

$$P(A_1 \cap A_2 \cap A_3 \cap A_4) = \frac{4}{52} \times \frac{4}{51} \times \frac{4}{50} \times \frac{3}{49} = 0.00003 \qquad ▲$$

Example 2.4 For the events A_1, A_2, and A_3, $P(A_1) = 0.5$, $P(A_2 | A_1) = 0.6$, $P(A_1 | A_2) = 0.8$, and $P(A_3 | A_1 \cap A_2) = 0.4$. Find $P(A_1 \cap A_3 | A_2)$.

Since $P(A_1 \cap A_3 | A_2) = P(A_1 \cap A_2 \cap A_3)/P(A_2)$, we have to compute $P(A_1 \cap A_2 \cap A_3)$ and $P(A_2)$. By (2.4), $P(A_1 \cap A_2 \cap A_3) = 0.5 \times 0.6 \times 0.4 = 0.12$. For finding $P(A_2)$, we evaluate $P(A_1 \cap A_2)$ by (2.3) in two different ways. First, $P(A_1 \cap A_2) = P(A_1)P(A_2 | A_1) = 0.5 \times 0.6 = 0.3$. Next, $P(A_1 \cap A_2) = P(A_2)P(A_1 | A_2) = 0.8 P(A_2)$; hence $0.8P(A_2) = 0.3$, from which $P(A_2) = 3/8$. Upon collecting the appropriate terms, we get $P(A_1 \cap A_3 | A_2) = 0.12/(3/8) = 0.32$.   ▲

Example 2.5 Find $P(A \cup B)$ if $P(A) = 0.6$, $P(B) = 0.5$, and $P(A | B) = 0.4$.

We know from Theorem 1.4 that $P(A \cup B) = P(A) + P(B) - P(A \cap B)$. Now, by (2.3), $P(A \cap B) = 0.4 \times 0.5 = 0.2$, and thus $P(A \cup B) = 0.6 + 0.5 - 0.2 = 0.9$.   ▲

2.3 THE TOTAL PROBABILITY RULE AND BAYES' THEOREM

One of the most powerful computation method of probability is the total probability rule, to which the present section is devoted. Before formulating it in its generality, let us look at the following example.

Example 2.6 There are two red and five white balls in urn 1 and three red and four white balls in urn 2. One ball is picked at random from urn 1 and, without checking its color, it is placed into urn 2; then one ball is picked at random from urn 2. Find the probability that the second ball chosen is red.

Let A be the event that the second ball drawn is red. It would be easy to evaluate $P(A)$ if we knew the color of the first choice. Therefore, we introduce the events

$$C_1 = \{\text{first is red}\}, \qquad C_2 = \{\text{first is white}\}$$

and we try to find $P(A)$ through the conditional probabilities $P(A | C_1)$ and $P(A | C_2)$. First note that

(2.6) $\qquad\qquad C_1 \cup C_2 = \Omega \qquad$ and $\qquad C_1 \cap C_2 = \emptyset$.

Consequently, if we write

$$A = A \cap \Omega = A \cap (C_1 \cup C_2) = (A \cap C_1) \cup (A \cap C_2)$$

the terms of the union on the right-hand side are mutually exclusive, and thus

$$P(A) = P(A \cap C_1) + P(A \cap C_2)$$

which, from (2.3), can also be written as

(2.7) $P(A) = P(C_1)P(A \mid C_1) + P(C_2)P(A \mid C_2)$

The terms of (2.7) are easy to compute; we have

$$P(A) = \frac{2}{7} \times \frac{4}{8} + \frac{5}{7} \times \frac{3}{8} = \frac{23}{56}$$ ▲

Note that the key formula (2.7) in the solution of the preceding example was derived from the properties in (2.6). We thus generalize the method of solution in Example 2.6 as follows.

Definition 2.2   A finite or denumerably infinite sequence C_1, C_2, ... of events is called a *complete system* (of events), or a *partition* of Ω, if $P(C_j) > 0$ for each j, $C_i \cap C_j = \emptyset$ for all $i \neq j$, and $P(\bigcup C_j) = 1$, where the union is taken over all j.

Theorem 2.3   (**The Total Probability Rule**)   Let C_1, C_2, ... be a complete system of events. Then, for an arbitrary event A,

(2.8) $P(A) = \sum P(A \mid C_j)P(C_j)$

where the summation is over all j.

Proof: Write $\Omega = B \bigcup (\bigcup C_j)$, where B and the C_j are mutually exclusive (B is possibly empty). Hence

$$1 = P(\Omega) = P(B) + P(\bigcup C_j) = P(B) + 1$$

and thus $P(B) = 0$. Now,

$$A = A \cap \Omega = A \cap [B \bigcup (\bigcup_j C_j)] = (A \cap B) \bigcup [\bigcup_j (A \cap C_j)]$$

from which

(2.9) $P(A) = P(A \cap B) + \sum_j P(A \cap C_j)$

The term $P(A \cap B) = 0$ because $A \cap B \subset B$ implies that $0 \leq P(A \cap B) \leq P(B) = 0$. Therefore, upon applying (2.3) to the terms $P(A \cap C_j)$, (2.9) leads to (2.8). The theorem is proved.   ▲

Example 2.7   At a factory, three machines M_1, M_2, and M_3 produce the same type of product. M_1 produces 40% of all products, M_2 produces 35%, and M_3 produces 25%. Out of those produced by M_1, M_2, and M_3, 5%, 3%, and 4%, respectively, are defective. The products are mixed in a storage room, from which one is picked at random, and sold. What is the probability that the product sold is defective? What is the conditional probability that the product sold came from M_1, given that it was found defective?

Let A be the event that a product in the storage room is defective. Furthermore, let C_1, C_2, and C_3 be the events that a product in the storage room came from M_1, M_2, and M_3, respectively. Then C_1, C_2, and C_3 form a complete system of events with $P(C_1) = 0.4, P(C_2) = 0.35,$ and $P(C_3) = 0.25$. We also know that $P(A \mid C_1) = 0.05$, $P(A \mid C_2) = 0.03$, and $P(A \mid C_3) = 0.04$. Hence, by the total probability rule,

$$P(A) = 0.05 \times 0.4 + 0.03 \times 0.35 + 0.04 \times 0.25 = 0.045$$

We now compute the conditional probability $P(C_1 \mid A)$, which by definition, equals $P(C_1 \cap A)/P(A)$. Since

$$P(C_1 \cap A) = P(A \mid C_1)P(C_1) = 0.05 \times 0.4 = 0.02$$

we get $P(C_1 \mid A) = 0.02/0.045 = 0.44$ (i.e., M_1 produces 44% of the defective products compared with its share of 40% of all products). ▲

Note that even if we were interested in the second question only, we should have computed $P(A)$ by the total probability rule. Such an application of the total probability rule is known as *Bayes' theorem*:

Theorem 2.4   Let C_1, C_2, ... be a complete system of events. Then, for an arbitrary event A with $P(A) > 0$,

$$(2.10) \qquad P(C_k \mid A) = \frac{P(A \mid C_k)P(C_k)}{\sum P(A \mid C_j)P(C_j)} \qquad k = 1, 2, \ldots$$

Proof: The proof is a simple combination of three previously established formulas. First, by the definition of conditional probabilities, $P(C_k \mid A) = P(C_k \cap A)/P(A)$. Next, by (2.3), $P(C_k \cap A) = P(A \mid C_k)P(C_k)$, and finally, the denominator $P(A)$ can be computed by the total probability rule. ▲

Could you get the correct answer to the question in the following example without Bayes' theorem?

Example 2.8   A box contains three disks; one is colored red on both sides, one is green on both sides, and the third one is red on one side and green on the other. One disk is picked at random from the box and the side that is up is seen to be red. What is the probability that the other side of this disk is also red?

Let $C_1 = \{$disk with two red sides is picked$\}$, $C_2 = \{$disk with two green sides is picked$\}$, and $C_3 = \{$disk with one red and one green side is picked$\}$. Further, let $A = \{$a disk shows a red side up$\}$. We want to compute $P(C_1 \mid A)$. Since C_1, C_2, and C_3 form a complete system of events, we can apply Bayes' theorem. Evidently, $P(C_j) = 1/3$ for $1 \le j \le 3$. Now, $P(A \mid C_1) = 1$, $P(A \mid C_2) = 0$, and $P(A \mid C_3) = 1/2$, and thus (2.10) with $k = 1$ yields

$$P(C_1 \mid A) = \frac{1/3}{1/3 + 0 + 1/2 \times 1/3} = \frac{2}{3} \qquad \blacktriangle$$

Bayes' theorem can be viewed as a mechanism for "correcting" previously "known" probabilities (usually just assumed values or roughly estimated) of some events as new information becomes available in connection with these events. The next example serves to show this.

Example 2.9   Out of every 20 units produced by an employee, M can be sold at a premium, where M varies, but the employee has been rewarded by management according to the following formula: $M = 12$ in 30% of the cases, $M = 14$ in 40% of the cases, and $M = 16$ in 30% of the cases. As a check, from a batch of 20 products, 2 are selected at random and both are found to be of premium quality. What would now be the proper formula for the reward?

Let $C_1 = \{M = 12\}$, $C_2 = \{M = 14\}$, and $C_3 = \{M = 16\}$. Because the C_j are mutually exclusive and by the assumption of management, $P(C_1) = 0.3$, $P(C_2) = 0.4$, and $P(C_3) = 0.3$ [i.e., $P(C_1 \cup C_2 \cup C_3) = 1$], they form a complete system. Thus, letting A denote the event that when selecting 2 out of 20 products by the employee in question, both are of premium quality, the desired probabilities $P(C_j \mid A)$, $1 \le j \le 3$, can be computed by Bayes' theorem. For this, we need $P(A \mid C_j)$, $1 \le j \le 3$. These conditional probabilities, however, are simply hypergeometric probabilities (Theorem 1.7), where the value of M is being specified by the condition C_j. As a matter of fact [recall that $\binom{n}{0} = 1$ for all $n \ge 1$],

$$P(A \mid C_1) = \frac{\binom{12}{2}}{\binom{20}{2}} = \frac{66}{190}, \quad P(A \mid C_2) = \frac{\binom{14}{2}}{\binom{20}{2}} = \frac{91}{190}, \quad P(A \mid C_3) = \frac{\binom{16}{2}}{\binom{20}{2}} = \frac{120}{190}$$

Substituting these values into (2.10), we get

$$P(C_1 \mid A) = \frac{66 \times 0.3}{66 \times 0.3 + 91 \times 0.4 + 120 \times 0.3} = 0.21$$

$$P(C_2 \mid A) = \frac{91 \times 0.4}{66 \times 0.3 + 91 \times 0.4 + 120 \times 0.3} = 0.39$$

and $P(C_3 \mid A) = 1 - 0.21 - 0.39 = 0.4$. That is, the corrected formula is 21%, 39%, and 40%, respectively, for $M = 12$, 14, and 16. ▲

2.4   THE INDEPENDENCE OF TWO EVENTS

For two events A and B with $P(B) > 0$, the effect of the occurrence of B on A is measured by the conditional probability $P(A \mid B)$. Evidently, if

(2.11a) $$P(A \mid B) = P(A)$$

then B has no effect on A, and thus it is natural to say that in this case, B is independent of A. Now, if $P(A) > 0$, the same reasoning leads to saying that A is independent of B if

(2.11b) $$P(B \mid A) = P(B)$$

The two equations (2.11a, b) seem to be different. However, upon replacing the conditional probabilities in these equations by their definition [see (2.2)], we arrive at the equation

(2.12) $$P(A \cap B) = P(A)P(B)$$

which is symmetric in A and B. This now explains the following definition.

Definition 2.3   Two events A and B are said to be *independent* if (2.12) holds.

Note that we dropped the requirement of $P(A) > 0$ or $P(B) > 0$. In fact, (2.12) always holds if one of $P(A)$ and $P(B)$ equals zero [both sides of (2.12) become zero; why?], and thus (2.12) is slightly different from (2.11a, b). But because events of zero probability are of little interest to us, we can say that for all practical purposes, (2.12) represents the same kind of independence as (2.11a, b).

Note also that if both $P(A) > 0$ and $P(B) > 0$ and if A and B are mutually exclusive, they cannot be independent (i.e., they are dependent). In such cases, the left-hand side of (2.12) is zero, but the right-hand side is positive.

Let us now look at some examples.

Example 2.10   The faces of each of a green (regular) die and a black (regular) die are marked by the numbers 1 through 6. Roll these two dice together, and let x and y be the numbers on the green die and the black die, respectively. Show that the events

(i) $A = \{x = 2\}$ and $B = \{y = 5\}$ are independent;
(ii) $A = \{x = 2\}$ and $C = \{x + y = 7\}$ are independent;
but
(iii) $A = \{x = 2\}$ and $D = \{x + y = 6\}$ are dependent.

The experiment of rolling two distinguishable dice has 36 possible outcomes; that is, there are six possibilities for x and six for y. Out of all outcomes, six are favorable to A ($x = 2$ and y can be any one of its six values) and six are favorable to B. Hence $P(A) = P(B) = 6/36 = 1/6$. Now, there is only one outcome favorable to $A \cap B$; thus $P(A \cap B) = 1/36$. By substituting into (2.12), we see that it holds; therefore, A and B are independent. Turning to part (ii), we find that C has six favorable outcomes: $(1, 6), (2, 5), (3, 4), (4, 3), (5, 2)$, and $(6, 1)$, where an outcome is written in the form (x, y). Furthermore, there is again only one outcome favorable to $A \cap C$. Thus $P(C) = 6/36 = 1/6$ and $P(A \cap C) = 1/36$, and (2.12) holds again, meaning that A and C are independent. But A and D in part (iii) are dependent, because $P(A \cap D) = 1/36$ and $P(A) = 1/6$; furthermore, there being only five outcomes favorable to D, $P(D) = 5/36$, and thus (2.12) fails to hold.   ▲

Although the values 2 and 5 had no significance in the definition of A and B, respectively, in the preceding example, the value 7 evidently had significance for $x + y$. It is interesting to note that both A and C formally depend on x, yet they are independent from the point of view of probability theory.

Note that the relation of A and C and that of A and D look similar, but in view of the conclusion of the example, there should be a basic difference between them. Such a difference can be recognized when we look not only at the relation between A and C (or A and D), but also at their relation to the original experiment. That is, in the whole experiment (of rolling the two dice) there are 36 outcomes, out of which six are favorable to A. Now, if we compare A and C, we see that there are six outcomes constituting C, out of which one is favorable to A; that is, the "position" of A within the whole experiment is the same as in C (in both cases, 1/6 of the elements belong to A). This led to the independence of A and C. However, if we compare A and D, we find that, out of the five outcomes favorable to D, one is favorable to A, that is, the "position" of A in the whole experiment (1/6 of the cases) is different from that in D (1/5 of the cases).

Example 2.11   From a lot of M type I and $T - M$ type II items, pick two at random, once without replacement, and once with replacement. Let A and B, respectively, be the events that the first and the second items drawn are type I. Let us show that A and B are dependent if selection is without replacement, and A and B are independent in the case of selection with replacement.

Even though we saw in Exercise 31 of Chapter 1 that $P(A) = P(B) = M/T$, whether selection is with or without replacement, let us reestablish this fact with the help of the total probability rule. It is, of course, clear that $P(A) = M/T$. For calculating $P(B)$, note that A and A^c form a complete system of events. Thus, by the total probability rule [see (2.8)],

$$(2.13) \qquad P(B) = P(B \mid A)\,P(A) + P(B \mid A^c)P(A^c)$$

Now, when drawing without replacement,

$$(2.14) \qquad P(B \mid A) = \frac{M-1}{T-1}, \qquad P(B \mid A^c) = \frac{M}{T-1}$$

from which, by substitution into (2.13), we have

$$P(B) = \frac{M-1}{T-1}\frac{M}{T} + \frac{M}{T-1}\frac{T-M}{T} = \frac{M}{T}$$

The last result, combined with (2.14), implies that $P(B \mid A) \neq P(B)$. That is, (2.11b) fails, and thus so does (2.12); consequently, A and B are dependent.

In the case of selection with replacement, $P(B \mid A) = M/T$, and one can compute it either directly or from (2.13), $P(B) = M/T$. Since $P(B) = P(B \mid A)$, A and B are independent.   ▲

Let us pause for a simple but significant theoretical result.

Theorem 2.5   Let A and B be independent events. Then so are A^c and B, A and B^c, and A^c and B^c.

Proof: By assumption, $P(A \cap B) = P(A)P(B)$. Our aim is to show that

$$(2.15) \qquad P(A^c \cap B) = P(A^c)P(B)$$

We start with the formula (see Theorem 1.4.2)

$$P(A^c \cap B) = P(B) - P(A \cap B)$$

which, by the assumption on A and B, becomes

$$P(A^c \cap B) = P(B) - P(A)P(B) = P(B)[1 - P(A)]$$

We know from Chapter 1 that $1 - P(A) = P(A^c)$, and thus the last equation reduces to (2.15), which was to be proved.

Because the argument is symmetric in A and B, we actually proved that both (A^c and B) and (A and B^c) are independent. From the independence of A^c and B, however, we can now conclude that A^c and B^c are independent as well, which completes the proof.   ▲

By the appropriate formulas of Chapter 1, let us solve the following problem.

Example 2.12   Let the events A and B be independent with $P(A) = 0.6$ and $P(B) = 0.3$. Find $P(A \cup B^c)$.

We know that $P(A \cup B^c) = P(A) + P(B^c) - P(A \cap B^c)$. By Theorem 2.5, $P(A \cap B^c) = P(A)P(B^c)$ (independence), and thus since $P(B^c) = 1 - P(B) = 0.7$,

$$P(A \cup B^c) = 0.6 + 0.7 - 0.6 \times 0.7 = 0.88$$ ▲

2.5   EXTENSIONS OF THE CONCEPT OF INDEPENDENCE

We would like to extend the concept of independence to more than two events. Let us first go back to Example 2.10. There we considered three events A, B, and C in connection with rolling two distinguishable dice, any two of which were shown to be independent. In spite of this, we wish to point out that it is unreasonable to consider them independent. Let us recall that denoting by (x, y) the numbers on the two dice, $A = \{x = 2\}$, $B = \{y = 5\}$, and $C = \{x + y = 7\}$. Now, consider $A \cap B$ as a single event. In any reasonable definition of independence of A, B, and C we must require that $A \cap B$ be independent of C. This, however, is not the case. That is, as we found in Example 2.10, $P(A \cap B) = 1/36$ and $P(C) = 1/6$; furthermore, $(A \cap B) \cap C = A \cap B$, and thus $P((A \cap B) \cap C) = 1/36$. Consequently,

$$P((A \cap B) \cap C) \neq P(A \cap B)P(C)$$

that is, $A \cap B$ and C are dependent. This example shows that when defining the independence of several events, we have to require more than the independence of all pairs.

Definition 2.4   We say that the events A_1, A_2, ..., A_n are independent if for every pair of disjoint subsets $(i_1, i_2, ..., i_k)$ and $(j_1, j_2, ..., j_l)$ of the subscripts $1, 2, ..., n$, the events

$$(2.16) \qquad A = A_{i_1} \cap A_{i_2} \cap \cdots A_{i_k}, \qquad B = A_{j_1} \cap A_{j_2} \cap \cdots \cap A_{j_l}$$

are independent, that is, if they satisfy (2.12). An equivalent form of the stated requirement is that for every $2 \leq m \leq n$, and every subset $(s_1, s_2, \ldots, s_m)$ of the subscripts 1, 2, ..., n,

$$(2.17) \qquad P(A_{s_1} \cap A_{s_2} \cap \cdots \cap A_{s_m}) = P(A_{s_1})P(A_{s_2}) \cdots P(A_{s_m})$$

The fact that the independence of the events of (2.16) is equivalent to (2.17) can easily be seen by induction. First take $k = t = 1$ in (2.16); then the independence of A and B is exactly (2.17) with $m = 2$. Now, taking $k = 2$ and $t = 1$ in (2.16), the independence of A and B means that

$$(2.17a) \qquad P(A_{i_1} \cap A_{i_2} \cap A_{j_1}) = P(A_{i_1} \cap A_{i_2})P(A_{j_1})$$

But since we have just established that with $m = 2$, (2.17) is applicable, (2.17a) becomes (2.17) with $m = 3$. That is, step by step, we can increase the value of m in (2.17) on the base of the assumption of the independence of A and B in (2.16). In other words, the independence of A and B in (2.16) implies (2.17). The converse of this statement can be established in a similar manner; that is, the two forms of the definition of independence are equivalent.

Note that (2.17) represents a very fast increasing number of equations as n increases. As a matter of fact, for fixed m, there are $\binom{n}{m}$ choices of the subscripts $s_1, s_2, \ldots, s_m$, and thus (2.17) represents

$$\binom{n}{2} + \binom{n}{3} + \cdots + \binom{n}{n} = 2^n - \binom{n}{1} - \binom{n}{0} = 2^n - n - 1$$

equations. For example, if $n = 10$, $2^n - n - 1 = 1013$, which means that in order to check whether 10 events are independent, we have to check the validity of 1,013 equations. This is not a real problem, however, because in theoretical discussions, one can usually check all the equations of (2.17) by a single argument; in practice, on the other hand, independence is usually an assumption.

Example 2.13   Pick a number x at random out of the integers 1 through 30. Let A be the event that x is even, B that x is divisible by 3, and C that x is divisible by 5. Let us show that the events A, B, and C are independent.

By taking the ratio of the number of favorable cases to the number of all possible choices (30), we get $P(A) = 15/30 = 1/2$, $P(B) = 10/30 = 1/3$, and $P(C) = 6/30 = 1/5$. Now, since $A \cap B$ means that x is divisible by 6, $P(A \cap B) = 5/30 = 1/6$, and thus $P(A \cap B) = P(A)P(B)$. Similarly,

$$P(A \cap C) = P(x \text{ is divisible by } 10) = \frac{3}{30} = \frac{1}{10} = P(A)P(C)$$

$$P(B \cap C) = P(x \text{ is divisible by } 15) = \frac{2}{30} = \frac{1}{15} = P(B)P(C)$$

and

$$P(A \cap B \cap C) = P(x \text{ is divisible by } 30) = \frac{1}{30} = P(A)P(B)P(C)$$

that is, (2.17) holds, and thus A, B, and C are independent.   ▲

Example 2.14   From a lot of M type I and $T - M$ type II items, let us select t items at random with replacement. Let A_1, A_2, ..., A_t be the events that the first, second, ..., tth selection, respectively, resulted in type I item. Then the events A_1, A_2, ..., A_t are independent.

We have to show that (2.17) holds. The intersection

(2.18) $A_{s_1} \cap A_{s_2} \cap \cdots \cap A_{s_m}$

means that in m specific positions type I items are chosen, and in the remaining $t - m$ positions any of the items (regardless of type) can come into account. Hence the number of favorable cases to (2.18) is $M^m T^{t-m}$ (selection is with replacement). On the other hand, the number of all possibilities in selecting t out of T with replacement is T^t, which we can also write as $T^m T^{t-m}$. Consequently,

$$P(A_{s_1} \cap A_{s_2} \cap \cdots \cap A_{s_m}) = \left(\frac{M}{T}\right)^m$$

which is exactly (2.17), because $P(A_j) = M/T$ for all j.   ▲

We have seen that the independence of two events is not affected if one or both are replaced by their complements. This remains valid for arbitrary number of events.

Theorem 2.6   If the events A_1, A_2, ..., A_n are independent, then so are A_1^c, A_2, ..., A_n. In fact, independent events remain independent if any number of them are replaced by their complements.

Proof: For the first part we have to prove that (2.17) implies that similar equations remain to hold if A_1 is replaced by A_1^c. Evidently, if all $s_j > 1$ in (2.17), the replacement of A_1 by A_1^c has no effect. Therefore, let $s_1 = 1$. Then writing $B = A_{s_2} \cap A_{s_3} \cap \cdots \cap A_{s_m}$, (2.17) becomes $P(A_1 \cap B) = P(A_1)P(B)$. But, by Theorem 2.5, this last equation implies the independence of A_1^c and B, which means that (2.17) remains to hold when A_1^c replaces A_1.

The second part of the theorem is an immediate consequence of the first part. That is, the first part can evidently be read as saying that independent events remain independent if any one of them is replaced by its complement. Now, if we repeat this statement k times, we get that independence is not affected if any k of n independent events are replaced by their complements. The theorem is proved. ▲

Example 2.15 Each of the four engines of an aircraft functions with probability 0.98, and they function independently of each other. The aircraft is considered safe if at least two of its engines function. Let us find the probability that such an aircraft is safe as far as its engines are concerned.

Let A_j be the event that the jth engine functions. If S signifies the event that the aircraft is safe, then S^c is the event that either all four engines have failed or that one functions and three have failed. These two cases are mutually exclusive, and thus, by symmetry,

$$P(S^c) = P(A_1^c \cap A_2^c \cap A_3^c \cap A_4^c) + 4P(A_1 \cap A_2^c \cap A_3^c \cap A_4^c)$$

Because the A_j are assumed independent, Theorem 2.6 gives

$$P(S^c) = (1 - 0.98)^4 + 4 \times 0.98 \times (1 - 0.98)^3 = 3.14 \times 10^{-5}$$

and thus $P(S) = 0.9999686$. ▲

We now give two additional extensions of the concept of independence.

Definition 2.5 We call an infinite sequence A_1, A_2, ... of events independent if, for every finite $n \geq 2$, the events A_1, A_2, ..., A_n are independent.

Definition 2.6 Successive experiments are called independent if the events A_1, A_2, ... are independent whenever A_j is an event whose occurrence is determined by the jth experiment.

We can thus speak of n independent repetitions of an experiment. This means that we perform the same experiment repeatedly n times; that is, the set of possible outcomes is the same every time, and if A is an event in connection with the experiment, its probability does not change from experiment to experiment. Furthermore, the results in any of these repetitions have no influence on the others. The requirement of repeating experiments independently of each other imposes a strict rule on the experimenter; in practice, however, it is rare that the assumption of independence could be checked (e.g., in such cases as the assumption of independence of the behavior of customers—their arrivals, service requirements, and others). In some cases it is simply impossible to repeat the same experiment independently of the others. That is, if repetition of the experiment involves selection of items from a lot, the selection should be with replacement in order that the repetitions be independent (Examples 2.11 and 2.14). Now, if the experiment involves both selection and the destruction of the selected item in a test (such as the random length of time that it takes to burn out an electric light bulb), the item cannot be placed back into the lot. However, if relatively few items are to be selected from a large lot, the independence requirement is only slightly violated, and what we say later for independently repeated trials will give a good approximation to all these cases.

2.6   A RETURN TO THE RELATION OF THE RELATIVE FREQUENCY AND PROBABILITY

We are no in the position to establish mathematical relations between the relative frequency of an event in independent repetitions of an experiment and its probability introduced in an abstract way through the axioms. We first prove the following exact formula, from which asymptotic relations will be deduced.

Theorem 2.7   Let A be an event in connection with a random experiment, and let $P(A) = p$. Let us repeat the experiment n times independently of each other. Let $k_A(n)$ be the frequency of A in these repetitions (i.e., the number of times when A occurred). Then

$$(2.19) \qquad P(k_A(n) = m) = \binom{n}{m} p^m (1-p)^{n-m} \qquad 0 \le m \le n$$

Remark   We have established this formula in a special case, the model of selection with replacement (Section 1.8). In Example 2.14 we have seen that that selection model does correspond to independent repetitions of selecting one item from the lot. Just as in Section 1.8, we call the probabilities of (2.19) the *binomial probabilities* (with parameters n and p).

Proof of Theorem 2.7:   Put $B_m = \{k_A(n) = m\}$. Let A_j, $1 \le j \le n$, be the event that A occurs in the jth repetition of the experiment. Hence A_1, A_2, ..., A_n are independent and $P(A_j) = p$ for all j. Now, B_m occurs if, and only if, exactly m of the A_j occur; in symbols,

$$(2.20) \qquad B_m = \bigcup (A_{j_1} \cap A_{j_2} \cap \cdots \cap A_{j_m} \cap A_{j_{m+1}}^c \cap \cdots \cap A_{j_n}^c)$$

where the union is taken over all possible choices of the subscripts $1 \le j_1 < j_2 < \cdots < j_m \le n$. The terms of the union above are mutually exclusive, and their number is the number of ways of choosing m out of n, that is, $\binom{n}{m}$. Therefore, since [by the independence of the A_j and by $P(A_j) = p$]

$$P(A_{j_1} \cap A_{j_2} \cap \cdots \cap A_{j_m} \cap A^c_{j_{m+1}} \cap \cdots \cap A^c_{j_n}) = p^m(1-p)^{n-m}$$

that is, the same value for all proper choices of the subscripts j_t, (2.20) yields (2.19). The proof is completed.   ▲

Note that the events $\{k_A(n) = m\}$, $0 \le m \le n$, are mutually exclusive, and one of them always occurs; therefore, the equation

$$(2.21) \qquad \sum_{m=0}^{n} \binom{n}{m} p^m (1-p)^{n-m} = 1$$

must hold. Now, if $a > 0$ and $b > 0$ are arbitrary, then upon setting $p = a/(a+b)$, from which $1 - p = b/(a+b)$, (2.21) becomes

$$(2.22) \qquad \sum_{m=0}^{n} \binom{n}{m} a^m b^{n-m} = (a+b)^n$$

a familiar formula from algebra. It is remarkable that we established (2.22) by a purely probabilistic argument.

Example 2.16   Roll a regular die six times, each independently of the others. Find the probability that exactly m times, $0 \le m \le 6$, the number 6 turns up.

Let A be the event that a 6 turns up when rolling a regular die. Then $p = P(A) = 1/6$. With the notations of Theorem 2.7, we want to evaluate $P(k_A(6) = m) = p_m$, $0 \le m \le 6$. By (2.19),

$$p_0 = \binom{6}{0}\left(\frac{1}{6}\right)^0\left(\frac{5}{6}\right)^6 = \left(\frac{5}{6}\right)^6, \qquad p_1 = \binom{6}{1}\left(\frac{1}{6}\right)^1\left(\frac{5}{6}\right)^5 = \frac{6 \times 5^5}{6^6}$$

$$p_2 = \binom{6}{2}\left(\frac{1}{6}\right)^2\left(\frac{5}{6}\right)^4 = \frac{3 \times 5^5}{6^6}, \qquad p_3 = \binom{6}{3}\left(\frac{1}{6}\right)^3\left(\frac{5}{6}\right)^3 = \frac{3 \times 4 \times 5^4}{6^6}$$

$$p_4 = \binom{6}{4}\left(\frac{1}{6}\right)^4\left(\frac{5}{6}\right)^2 = \frac{3 \times 5^3}{6^6}, \qquad p_5 = \binom{6}{5}\left(\frac{1}{6}\right)^5\left(\frac{5}{6}\right)^1 = \frac{6 \times 5}{6^6}$$

and

$$p_6 = \binom{6}{6}\left(\frac{1}{6}\right)^6\left(\frac{5}{6}\right)^0 = \frac{1}{6^6} \qquad\qquad ▲$$

It is immediately seen that p_1 is the largest, and in fact $p_0 < p_1$ and $p_1 > p_2 > p_3 > p_4 > p_5 > p_6$. The reason for the special role of p_1 is that with $k_A(6) = 1$, the relative frequency $k_A(6)/6$ equals $p = 1/6$. That is, the relative frequency in the most likely case (p_1 is the largest among all p_m) equals the probability. This observation is generally true in the following slightly modified form.

Theorem 2.8   Among the binomial probabilities (2.19) there is a largest one [i.e., there is a most likely value of the frequency $k_A(n)$]. This most likely relative frequency satisfies the inequality

(2.23)
$$\left| \frac{k_A(n)}{n} - p \right| < \frac{1}{n}$$

Proof: Set $p_m = P(k_A(n) = m)$. Then, from (2.19),

$$\frac{p_{m+1}}{p_m} = \frac{\dfrac{n!}{(m+1)!\,(n-m-1)!}\, p^{m+1}(1-p)^{n-m-1}}{\dfrac{n!}{m!\,(n-m)!}\, p^m(1-p)^{n-m}} = \frac{(n-m)p}{(m+1)(1-p)}$$

Hence $p_{m+1} > p_m$ if, and only if, $(n-m)p > (m+1)(1-p)$, or equivalently, if $m < np - (1-p)$. That is, up to the value $m < np - (1-p)$, $\{p_m\}$ is an increasing sequence, and if $m > np - (1-p)$, $\{p_m\}$ is decreasing. Consequently, the largest p_m is such that $|m - np| < 1 - p < 1$, which, when divided by n, becomes (2.23), which was to be proved.   ▲

Note that (2.23) reduces to $k_A(n)/n = p$ if np is an integer, which is the case in Example 2.16. In any case, (2.23) implies that, when n is large, the most likely relative frequency $k_A(n)/n$ is "very close" to $p = P(A)$. We can improve on this relation by dropping the restrictive requirement of closeness "in the most likely case." As a matter of fact, we now state two of the most important results of probability theory, in which we use the notations and assumptions of Theorem 2.7.

Theorem 2.9   (The Chebyshev Inequality)   For $n \geq 1$, and for every $\varepsilon > 0$,

(2.24)
$$P\left(\left| \frac{k_A(n)}{n} - p \right| \geq \varepsilon \right) \leq \frac{p(1-p)}{\varepsilon^2 n} \leq \frac{1}{4\varepsilon^2 n}$$

In particular, as $n \to +\infty$,

(2.25)
$$\lim P\left(\left| \frac{k_A(n)}{n} - p \right| < \varepsilon \right) = 1$$

Theorem 2.10 (The Normal Approximation) For $a < b$, and for $n \to +\infty$,

(2.26)
$$\lim P\left(a\left[\frac{p(1-p)}{n}\right]^{1/2} \leq \frac{k_A(n)}{n} - p \leq b\left[\frac{p(1-p)}{n}\right]^{1/2}\right) = N(b) - N(a)$$

where

(2.27)
$$N(x) = (2\pi)^{-1/2} \int_{-\infty}^{x} e^{-t^2/2} dt$$

whose values are given in Table 2.1. The limit relation (2.26) remains to hold if $a = -\infty$ and/or if $b = +\infty$, in which case the values $N(-\infty) = 0$ and $N(+\infty) = 1$ apply.

Before proceeding, let us record here that

(2.28)
$$N(x) = 1 - N(-x)$$

for all x. This and other properties of $N(x)$ are discussed in Section 2.7.

Remark The accuracy of the approximation in (2.26) depends not only on n but on p as well. As a rule of thumb we can use the fact that whenever $np(1-p) \geq 6$, (2.26) is applicable.

The proofs of these theorems are given in Sections 2.8 and 2.9. The rest of the present section is devoted to their applications.

Let us first emphasize the theoretical consequences of these results. The limit in (2.25) tells us that it is practically certain that, for sufficiently large n, the relative frequency $k_A(n)/n$ is as close to $p = P(A)$ as we wish (we choose $\varepsilon > 0$). This is what we were aiming at all along; the abstract concept of probability introduced through the axioms is the same value as the limiting form of the relative frequency. In other words, the computed probability p is practically the same value as the observed relative frequency $k_A(n)/n$ for large n. In this form, our deduction is not just theoretical; it gives us a rather practical rule of predicting $k_A(n)$ (as approximately np), and conversely, if there is no reasonable way of computing p, we can get its approximate value through observations (the relative frequency). The accuracy of this rule is expressed in (2.24) or (2.26).

Example 2.17 Roll a regular die $n = 1200$ times, each independently of the others. Evaluate the probability that the number k of times when a 6 turns up satisfies $180 < k < 220$.

Let A be the event that a 6 turns up when rolling a die. Then $k = k_A(1200)$ and $p = P(A) = 1/6$. Hence

$$P(180 < k < 220) = P\left(-\frac{20}{1200} < \frac{k}{n} - p < \frac{20}{1200}\right) = P\left(\left|\frac{k}{n} - p\right| < \frac{20}{1200}\right)$$

This expression is similar to the one in (2.24), except that the inequality in it is reversed. But an inequality can easily be reversed by turning to complements, and thus, by (2.24) with $\varepsilon = 20/1200 = 1/60$,

$$P(180 < k < 220) = 1 - P\left(\left|\frac{k}{n} - p\right| \geq \frac{1}{60}\right) \geq 1 - \frac{(1/6) \times (5/6)}{(1/60)^2 \times 1200} = 0.58$$

Let us also approximate this probability through (2.26). We again write

$$P(180 < k < 220) = P\left(-\frac{1}{60} < \frac{k}{n} - p < \frac{1}{60}\right)$$

where now a and b are to be computed from

$$-\frac{1}{60} = a\left[\frac{(1/6)(5/6)}{1200}\right]^{1/2} \quad \text{and} \quad \frac{1}{60} = b\left[\frac{(1/6)(5/6)}{1200}\right]^{1/2}$$

We get $a = -1.55$ and $b = 1.55$, and thus (2.26) gives

$$P(180 < k < 220) = N(1.55) - N(-1.55) = 2N(1.55) - 1 = 0.8788$$

where we utilized (2.28), and the last value was taken from Table 2.1. By comparing the result obtained by the Chebyshev inequality with the last one from the normal approximation, the advantage of the latter becomes clear.   ▲

Example 2.18 How many times do we have to roll a regular die in order to get

$$P\left(\left|\frac{k}{n} - \frac{1}{6}\right| < 0.02\right) \geq 0.95$$

where k is the frequency of a 6 in n rolls?

We answer this question by an appeal to the normal approximation. We first write

$$P\left(\left|\frac{k}{n} - \frac{1}{6}\right| < 0.02\right) = P\left(-0.02 < \frac{k}{n} - \frac{1}{6} < 0.02\right)$$

Setting $0.02 = b[(1/6)(5/6)/n]^{1/2}$, we get $b = -a = 0.12(n/5)^{1/2}$. Therefore, in view of $N(a) = 1 - N(-a) = 1 - N(b)$, (2.26) yields that n should be such that

$$2N\left(\frac{0.12}{\sqrt{5}}\sqrt{n}\right) - 1 \geq 0.95, \quad \text{that is,} \quad N\left(\frac{0.12}{\sqrt{5}}\sqrt{n}\right) \geq 0.975$$

Now, from Table 2.1, $N(1.96) = 0.975$, and thus, by the fact that $N(x)$ is an increasing function, we must have

$$0.12\left(\frac{n}{5}\right)^{1/2} \geq 1.96, \quad \text{that is,} \quad n \geq 5\left(\frac{1.96}{0.12}\right)^2 = 1333.8 \qquad \blacktriangle$$

Results like those in Examples 2.17 and 2.18 can be utilized in experimental tests for the accuracy of a die. One should, of course, get the same regularity for the frequency of each face, not just for that of 6.

Recall the remark following Example 1.27 in Section 1.8 concerning the close relation between the binomial and hypergeometric probabilities. Because of this close relation, the normal approximation is applicable to the hypergeometric probabilities as well under the following conditions. If n items are selected at random without replacement from a lot of T items, of which M are type I, then (2.26) is applicable with $p = M/T$, whenever both T and n are large, but n^2/T is approaching zero (p is assumed to be a fixed value). Here, of course, $k_A(n)$ signifies the number of type I items among those selected. We omit the proof of this statement, but those who go through the detailed proof of Theorem 2.10 in Section 2.9 can easily extend those arguments to the hypergeometric probabilities. Let us, however, look at an example in this connection.

Example 2.19   In an opinion poll of 1600 randomly selected citizens, 428 respond favorably to a specific government program. How accurately does this reflect the opinion of the whole population?

If T denotes the number of people in the population, of whom M favor the specified program, then $p = M/T$ is a fixed, but unknown, number. Our interest is to compare p with 428/1600. Assuming that T is considerably larger than $n^2 = 1600^2$, we can proceed with (2.26). We first decide that we want to determine the error in $k/n - p = 448/1600 - p$ with a certain degree of confidence, 95% say. Then, we determine $b = -a$ from $N(b) - N(a) = 2N(b) - 1 = 0.95$ [i.e., $N(b) = 0.975$]; we get from Table 2.1 the value $b = 1.96$. Hence, from (2.26), with 95% confidence,

$$|p - 0.28| < 1.96\frac{[p(1-p)]^{1/2}}{40}$$

(where we computed 448/1600 = 0.28) from which the uniformly applicable bound $p(1-p) \leq (1/2)(1-1/2) = 1/4$ gives $|p - 0.28| < 0.0245$, which can only slightly be improved if we compute $[p(1-p)]^{1/2}$ with a value close to 0.28 (e.g., with p around 0.3, the bound becomes 0.022). $\qquad \blacktriangle$

In the next two examples, we look at the relation $k_A(n)/n \sim p$ when p or $1 - p$ is supposed to be very small (matters of safety).

Example 2.20   Assume that the probability that a nuclear reactor operates safely in any given year is 0.999. Inspection and maintenance guarantee that this probability does not change over time, and assume that safe operations in subsequent years form independent events. For further increasing safety, an independent backup system is installed, whose safety level is the same as that of the main system. If the nation operates 200 nuclear reactors, can a breakdown be anticipated in one's lifetime?

If A is the event that a specific nuclear reactor breaks down in a given year, then $P(A) = 0.001$ if the backup system is not used, and $P(A) = 0.001 \times 0.001 = 10^{-6}$ if the backup system is considered as well (by the assumption of independence). Now, every year we have 200 repetitions for A, and thus in just 5 years we have $n = 1000$ repetitions of the "experiment" of operation of an atom reactor. Hence, if the backup system were not there (or would not come up), then the relation $k_A(n)/n \sim p$ implies that $k_A(1000) \sim 1000 \times 0.001 = 1$; that is, we could expect a breakdown in a 5-year period. However, if the backup system functions properly, then $k_A(n) \sim 10^{-6}n$ yields that, on average, a breakdown occurs only in every 5000 years.   ▲

Example 2.21   Assume that a screening test is 98% effective in detecting a certain type of cancer when present, and that the test shows the cancer for only 1% of those persons who do not suffer from it. Estimates show that 0.8% of the population has this form of cancer. With early detection of this type of cancer, an operation prevents its development, so every one whose test is positive undergoes surgery. Let us evaluate the probability of the event C that someone is free of the type of cancer in question, given that the test is positive for the patient (who thus undergoes unnecessary surgery).

Let A be the event that the test is positive. Then, by Bayes's theorem (Theorem 2.4),

$$P(C \mid A) = \frac{P(A \mid C)P(C)}{P(A \mid C)P(C) + P(A \mid C^c)P(C^c)}$$

$$= \frac{0.01 \times 0.992}{0.01 \times 0.992 + 0.98 \times 0.008} = 0.559$$

Consequently, about 56% of the people who have the surgery are actually free of the disease, even though the test appears to be highly effective in detecting the cancer when present and not showing it when it is not present.   ▲

2.7   THE STANDARD NORMAL DISTRIBUTION FUNCTION

In this section we establish several analytical properties of the function

$$(2.27) \qquad N(x) = (2\pi)^{-1/2} \int_{-\infty}^{x} e^{-t^2/2} \, dt$$

which we call the *standard normal distribution function*. Its significance has already been seen in Theorem 2.10; others will emerge in subsequent chapters.

Theorem 2.11   The function $N(x)$ satisfies the following properties: (i) increases in x, (ii) $N(-\infty) = 0$ and $N(+\infty) = 1$, and (iii) $N(-x) = 1 - N(x)$ for all x.

Proof: Note first that $N(x)$ is well defined for all finite x [i.e., the improper integral in (2.27) is finite; compare the integrand, for example, with e^t, $t < 0$]. Now, because $N'(x) = (2\pi)^{-1/2} e^{-x^2/2} > 0$, we know from calculus that $N(x)$ is strictly increasing. Hence part (i) is proved. For finding $N(+\infty)$, we actually compute $N^2(+\infty)$. We have

$$N^2(+\infty) = \left(\frac{1}{\sqrt{2\pi}} \int_{-\infty}^{+\infty} e^{-(1/2)x^2} dx \right) \left(\frac{1}{\sqrt{2\pi}} \int_{-\infty}^{+\infty} e^{-(1/2)y^2} dy \right)$$

$$= \frac{1}{2\pi} \int_{-\infty}^{+\infty} \int_{-\infty}^{+\infty} e^{-(1/2)(x^2 + y^2)} dx \, dy$$

We turn to polar coordinates (r, α), which leads to the substitution $x = r \cos \alpha$ and $y = r \sin \alpha$. From calculus,

$$dx \, dy = \left(\frac{\partial x}{\partial r} \frac{\partial y}{\partial \alpha} - \frac{\partial x}{\partial \alpha} \frac{\partial y}{\partial r} \right) dr \, d\alpha = r(\cos^2 \alpha + \sin^2 \alpha) \, dr \, d\alpha$$

$$= r \, dr \, d\alpha$$

and since every $0 \le r < +\infty$ and $0 \le \alpha \le 2\pi$ come into account,

$$N^2(+\infty) = \frac{1}{2\pi} \int_0^{2\pi} \left(\int_0^{+\infty} re^{-(1/2)r^2} dr \right) d\alpha = \frac{1}{2\pi} \int_0^{2\pi} \left[-e^{-(1/2)r^2} \right]_{r=0}^{+\infty} d\alpha$$

$$= \frac{1}{2\pi} \int_0^{2\pi} d\alpha = 1$$

from which $N(+\infty) = 1$. Because $N(x)$ is finite for all x, $N(-\infty)$ must be zero (both the upper and lower limits of the integral become $-\infty$). This completes the proof of part (ii).

Finally, by definition,

$$N(x) + N(-x) = (2\pi)^{-1/2} \int_{-\infty}^{x} e^{-t^2/2} \, dt + (2\pi)^{-1/2} \int_{-\infty}^{-x} e^{-t^2/2} \, dt$$

Table 2.1

z	0	1	2	3	4	5	6	7	8	9
.0	.5000	.5040	.5080	.5120	.5160	.5199	.5239	.5279	.5319	.5359
.1	.5398	.5438	.5478	.5517	.5557	.5596	.5636	.5675	.5714	.5753
.2	.5793	.5832	.5871	.5910	.5948	.5987	.6026	.6064	.6103	.6141
.3	.6179	.6217	.6255	.6293	.6331	.6368	.6406	.6443	.6480	.6517
.4	.6554	.6591	.6628	.6664	.6700	.6736	.6772	.6808	.6844	.6879
.5	.6915	.6950	.6985	.7019	.7054	.7088	.7123	.7157	.7190	.7224
.6	.7257	.7291	.7324	.7357	.7389	.7422	.7454	.7486	.7517	.7549
.7	.7580	.7611	.7642	.7673	.7704	.7734	.7764	.7794	.7823	.7852
.8	.7881	.7910	.7939	.7967	.7995	.8023	.8051	.8078	.8106	.8133
.9	.8159	.8186	.8212	.8238	.8264	.8289	.8315	.8340	.8365	.8389
1.0	.8413	.8438	.8461	.8485	.8508	.8531	.8554	.8577	.8599	.8621
1.1	.8643	.8665	.8686	.8708	.8729	.8749	.8770	.8790	.8810	.8830
1.2	.8849	.8869	.8888	.8907	.8925	.8944	.8962	.8980	.8997	.9015
1.3	.9032	.9049	.9066	.9082	.9099	.9115	.9131	.9147	.9162	.9177
1.4	.9192	.9207	.9222	.9236	.9251	.9265	.9279	.9292	.9306	.9319
1.5	.9332	.9345	.9357	.9370	.9382	.9394	.9406	.9418	.9429	.9441
1.6	.9452	.9463	.9474	.9484	.9495	.9505	.9515	.9525	.9535	.9545
1.7	.9554	.9564	.9573	.9582	.9591	.9599	.9608	.9616	.9625	.9633
1.8	.9641	.9649	.9656	.9664	.9671	.9678	.9686	.9693	.9699	.9706
1.9	.9713	.9719	.9726	.9732	.9738	.9744	.9750	.9756	.9761	.9767
2.0	.9772	.9778	.9783	.9788	.9793	.9798	.9803	.9808	.9813	.9817
2.1	.9821	.9826	.9830	.9834	.9838	.9842	.9846	.9850	.9854	.9857
2.2	.9861	.9865	.9868	.9872	.9875	.9878	.9881	.9884	.9887	.9890
2.3	.9893	.9896	.9899	.9902	.9904	.9906	.9909	.9911	.9913	.9916
2.4	.9918	.9920	.9922	.9924	.9927	.9929	.9931	.9933	.9934	.9936
2.5	.9938	.9940	.9942	.9943	.9945	.9947	.9948	.9949	.9951	.9952
2.6	.9953	.9954	.9955	.9957	.9958	.9960	.9961	.9962	.9963	.9964
2.7	.9965	.9966	.9967	.9968	.9969	.9970	.9971	.9972	.9973	.9974
2.8	.9975	.9975	.9976	.9976	.9977	.9978	.9979	.9980	.9980	.9981
2.9	.9981	.9982	.9982	.9983	.9984	.9984	.9985	.9985	.9986	.9986
3.0	.9987	.9987	.9987	.9988	.9988	.9989	.9989	.9989	.9990	.9990

Now, if we substitute $y = -t$ in the second integral, we get

$$N(x) + N(-x) = (2\pi)^{-1/2} \int_{-\infty}^{x} e^{-t^2/2}\, dt + (2\pi)^{-1/2} \int_{x}^{+\infty} e^{-y^2/2}\, dy$$

$$= (2\pi)^{-1/2} \int_{-\infty}^{+\infty} e^{-t^2/2}\, dt = 1$$

where, in the last equation, we utilized the result of part (ii). This completes the proof.   ▲

Because the function $y = \exp(-x^2)$ does not have an antiderivative among the elementary functions, $N(x)$ cannot be calculated directly. However, just as for the more familiar functions, such as $\sin x$ or $\log x$, tables are available for $N(x)$, and, in fact, it is easy to construct such a table. Note first that it is sufficient to make a table for $x > 0$ in view of part (iii) of Theorem 2.11. Now, when $N(x)$ has been computed, the equation

(2.28a) $N(x + h) = N(x) + h\{(2\pi)^{-1/2}e^{-[x + (1/2)h]^2/2}\}$

provides a very accurate value for $N(x + h)$ whenever h is "small." As a matter of fact, (2.28a) is arrived at through the definition of the integral as the limit of sums of the area of specific rectangles (called *Riemann sums*) (see Figure 2.1).

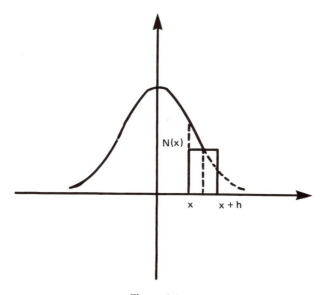

Figure 2.1

The values of $N(z)$ in Table 2.1 were computed by the author on a hand-held calculator (which has a key for each of 2π, square root, square, and e^x) by utilizing (2.28a) with $h = 0.01$ and the fact that $N(0) = 1/2$ [which follows, for example, from part (iii) of Theorem 2.11 with $x = 0$].

2.8   THE PROOF OF THE CHEBYSHEV INEQUALITY

We first prove two identities involving the binomial probabilities.

Lemma   For $n \geq 1$,

(2.29)
$$\sum_{m=0}^{n} m\binom{n}{m}p^m(1-p)^{n-m} = np$$

and

(2.30)
$$\sum_{m=0}^{n} (m-np)^2\binom{n}{m}p^m(1-p)^{n-m} = np(1-p)$$

Proof: First note that, for $m \geq 1$,

(2.31)
$$m\binom{n}{m} = n\binom{n-1}{m-1}$$

which can easily be obtained by writing both sides by means of factorials. Now, since the term corresponding to $m = 0$ in (2.29) is zero, we can actually start summation there with $m = 1$. Hence, from (2.31),

$$\sum_{m=0}^{n} m\binom{n}{m}p^m(1-p)^{n-m} = n\sum_{m=1}^{n}\binom{n-1}{m-1}p^m(1-p)^{n-m}$$

which indeed equals np, because if one removes p from behind the summation, the remaining sum is one in view of (2.21). This proves (2.29).

Let us turn to (2.30). Writing

(2.32)
$$(m-np)^2 = m(m-1) + (1-2np)m + (np)^2$$

we split (2.30) into three sums accordingly. For the sum containing $m(m-1)$, summation can start at $m = 2$. Hence, by applying (2.31) successively,

$$\sum_{m=0}^{n} m(m-1)\binom{n}{m}p^m(1-p)^{n-m} = n(n-1)\sum_{m=2}^{n}\binom{n-2}{m-2}p^m(1-p)^{n-m}$$

where, after removing p^2 from behind the summation sign, (2.21) can again be utilized to yield

(2.33)
$$\sum_{m=0}^{n} m(m-1)\binom{n}{m}p^m(1-p)^{n-m} = n(n-1)p^2$$

The second and third sums originating from the split in (2.32) are immediate from (2.29) and (2.21), respectively. We get

$$\sum_{m=0}^{n} (1 - 2np)m\binom{n}{m}p^m(1 - p)^{n-m} = (1 - 2np)np$$

and

$$\sum_{m=0}^{n} (np)^2\binom{n}{m}p^m(1 - p)^{n-m} = (np)^2$$

which, together with (2.32) and (2.33), yield for the left-hand side of (2.30) the value

$$n(n - 1)p^2 + (1 - 2np)np + (np)^2 = np - np^2 = np(1 - p)$$

which completes the proof of the lemma.   ▲

We can now turn to the proof of the Chebyshev inequality. First note that, putting $k = k_A(n)$,

$$P\left(\left|\frac{k}{n} - p\right| \geq \varepsilon\right) = P(|k - np| \geq \varepsilon n) = \sum{}^{*} P(k = m)$$

where Σ^* signifies summation over those values of m for which $|m - np| \geq \varepsilon n$. Hence

$$\varepsilon^2 n^2 P\left(\left|\frac{k}{n} - p\right| \geq \varepsilon\right) = \sum{}^{*} \varepsilon^2 n^2 P(k = m) \leq \sum{}^{*}(m - np)^2 P(k = m)$$

because, in Σ^*, every m satisfies $\varepsilon^2 n^2 \leq (m - np)^2$. This is increased further if we now sum for all m rather than only for those considered in Σ^*; we thus get

$$\varepsilon^2 n^2 P\left(\left|\frac{k}{n} - p\right| \geq \varepsilon\right) \leq \sum_{m=0}^{n} (m - np)^2 P(k = m) = np(1 - p)$$

the last equation being obtained by first applying the formula (2.19) for $P(k = m)$, and then the identity (2.30). If we divide the inequality above by $\varepsilon^2 n^2$, we get the first inequality of (2.24). The second inequality, on the other hand, is due to the simple fact that $p(1 - p)$ as a function of p on $(0, 1)$ is a parabola whose maximum is at $p = 1/2$, which equals $1/4$. This completes the proof of Theorem 2.9.   ▲

2.9 THE PROOF OF THEOREM 2.10

In this section we again start with a lemma whose content is known as *Stirling's formula*.

Lemma   For all integers $t \geq 2$,

(2.34) $$t! = Kt^{t+1/2}e^{-t}e^{c(t)/t} \qquad 0 < c(t) < 1$$

where K does not depend on t.

Remark   It will be shown later than $K = (2\pi)^{1/2}$.

Proof: We start with the logarithm of $t!$. One can easily verify the equation[+]

(2.35) $$\sum_{j=1}^{t} \log j = \int_{3/2}^{t+1/2} \log x \, dx + \int_{3/2}^{t+1/2} \left(x - [x] - \frac{1}{2}\right)\frac{1}{x}\,dx$$

where $[x]$ signifies the largest integer not exceeding x. We treat the two integrals of (2.35) separately. First, by integrating by parts, we get

(2.36) $$\int_{3/2}^{t+1/2} \log x \, dx = \left(t + \frac{1}{2}\right)\log\left(t + \frac{1}{2}\right) - t + c_1$$

where $c_1 = 1 - (3/2)\log(3/2)$, but whose value will be immaterial in what follows (except for the fact that it does not depend on t). In the second integral of (2.35), we carry out careful estimates. We write

$$\int_{3/2}^{t+1/2} \cdots = \int_{3/2}^{2} \cdots + \sum_{j=2}^{t-1}\left(\int_{j}^{j+1/2} \cdots + \int_{j+1/2}^{j+1} \cdots\right) + \int_{t}^{t+1/2} \cdots$$

where each of the ellipses is to be replaced by $(x - [x] - 1/2)/x\,dx$. Now, by substituting $y = j + 1/2 - x$ (note that $[x] = j$ below),

$$\int_{j}^{j+1/2}\left(x - [x] - \frac{1}{2}\right)\frac{1}{x}\,dx = -\int_{0}^{1/2}\frac{y}{j + 1/2 - y}\,dy$$

and if we substitute $y = x - j - 1/2$,

$$\int_{j+1/2}^{j+1}\left(x - [x] - \frac{1}{2}\right)\frac{1}{x}\,dx = \int_{0}^{1/2}\frac{y}{y + j + 1/2}\,dy$$

Upon combining the last three equations and denoting the second integral of the right-hand side of (2.35) by I_2, we have

$$I_2 = \int_{0}^{1/2}\frac{y}{y + 3/2}\,dy - \int_{0}^{1/2}\sum_{j=2}^{t-1}\frac{2y^2}{(j + 1/2)^2 - y^2}\,dy - \int_{0}^{1/2}\frac{y}{t + 1/2 - y}\,dy$$

If we change the sum in the middle term of I_2 by first extending the summation from $j = 2$ to infinity, and then subtracting a similar sum in which j runs from t to infinity, the only part in I_2 which depends on t becomes

---

[+]Throughout the book, $\log n$ signifies the natural logarithm of n.

$$\int_0^{1/2} \sum_{j=t}^{+\infty} \frac{2y^2}{(j+1/2)^2 - y^2}\, dy - \int_0^{1/2} \frac{y}{t+1/2-y}\, dy$$

Each of these integrals, however, can easily be estimated as a constant multiplied by the function $1/t$. That is, the absolute value of the second integral is increased if we decrease its denominator to t, which leads to the inequality

$$\int_0^{1/2} \frac{y}{t+1/2-y}\, dy < \frac{1}{t}\int_0^{1/2} y\, dy = \frac{1}{8t}$$

Similarly,

$$\int_0^{1/2} \sum_{j=t}^{+\infty} \frac{y^2}{(j+1/2)^2 - y^2}\, dy < \sum_{j=t}^{+\infty} \frac{1}{j^2+j}\int_0^{1/2} y^2\, dy = \frac{1}{24}\sum_{j=t}^{+\infty}\left(\frac{1}{j} - \frac{1}{j+1}\right)$$

which equals $1/(24t)$ (when the last sum is written out in detail, all terms except the very first cancel out). We thus established that $I_2 = c_2 + a(t)$ with $|a(t)| < c_3/t$, where c_2 and c_3 are constants.

To complete the proof of the lemma, we show that the right-hand side of (2.36) can also be written as

(2.36a) $\left(t+\frac{1}{2}\right)\log t - t + c_4 + b(t) \qquad |b(t)| < \frac{c_5}{t}$

where c_4 and c_5 are constants. As a matter of fact, since

$$\log\left(t+\frac{1}{2}\right) = \log t\left(1+\frac{1}{2t}\right) = \log t + \log\left(1+\frac{1}{2t}\right)$$

the inequality

$$|\log(1+x) - x| < \frac{1}{2}x^2 \qquad 0 < x < 1$$

immediately yields (2.36a) [to obtain this last inequality, simply integrate from zero to x the inequalities $1 - u < 1/(1+u) < 1$]. Now, (2.36), (2.36a), and the estimate on the second integral I_2 of (2.35) imply that

$$t! = \exp\left(\sum_{j=1}^t \log j\right) = \exp\left[\left(t+\frac{1}{2}\right)\log t - t + c_6 + \frac{c(t)}{t}\right]$$

$$= Kt^{t+1/2}e^{-t}e^{c(t)/t}$$

where $c_6 = c_2 + c_4$, $c(t)/t = a(t) + b(t)$, and $K = e^{c_6}$. The reader can easily determine these values of c_3 and c_5 from the preceding argument, which yields $0 < c(t) < 1$. This completes the proof of the lemma.   ▲

We now turn to the proof of Theorem 2.10. We want to achieve a good approximation to the sum

(2.37)

$$P\left(a\left[\frac{p(1-p)}{n}\right]^{1/2} < \frac{k_A(n)}{n} - p < b\left[\frac{p(1-p)}{n}\right]^{1/2}\right) = \sum' P(k_A(n) = m)$$

where $\sum'$ is summation over those values of m for which

$$a\left[\frac{p(1-p)}{n}\right]^{1/2} < \frac{m}{n} - p < b\left[\frac{p(1-p)}{n}\right]^{1/2}$$

or, equivalently,

(2.38) $np + a[np(1-p)]^{1/2} < m < np + b[np(1-p)]^{1/2}$

This means that, from (2.19), we need a good approximation to

(2.39) $$P(k_A(n) = m) = \frac{n!}{m!\,(n-m)!}p^m(1-p)^{n-m}$$

for all m satisfying (2.38). Since there are a constant times $n^{1/2}$ values of m in (2.38), the error term in approximating (2.39) should be such that even after multiplication by $n^{1/2}$, it should converge to zero as $n \to +\infty$.

    Let us write the values m in (2.38) as

(2.40) $m = np + y = np + x_m[np(1-p)]^{1/2}$ $a < x_m < b$

Note that $n - m = n(1-p) - y$. Now, utilizing (2.34) with $t = n$, m, and $n - m$ in (2.39), we get

(2.41)
$$P(k_A(n) = m) = \frac{\exp[c^*(n)/n]}{K[m(n-m)/n]^{1/2}}\left(1 + \frac{y}{np}\right)^{-m}\left[1 - \frac{y}{n(1-p)}\right]^{-(n-m)}$$

where we combined the terms

$$e^{c(n)/n}e^{-c(m)/m}e^{-c(n-m)/n-m} = e^{c^*(n)/n}$$

and thus $|c^*(n)|$ is bounded. For simplifying (2.41), we appeal to the Taylor expansions

$$e^x = 1 + x + \frac{x^2}{2} + \cdots$$

$$(1+x)^{-1/2} = 1 - \frac{1}{2}x + \cdots$$

and

$$\log(1+x) = x - \frac{x^2}{2} + \frac{x^3}{3} - \cdots$$

in which if we stop at a finite term, the error term is known to be of the magnitude of the next term in the expansion. With this in mind, $\exp[c^*(n)/n]$ becomes 1 plus an error of the magnitude of $1/n$ [because $c^*(n)$ is bounded]. Also, from

$$\frac{m(n-m)}{n} = (np+y)\left(1-p-\frac{y}{n}\right) = np(1-p)\left(1+\frac{y}{np}\right)\left[1-\frac{y}{n(1-p)}\right]$$

the second Taylor expansion above gives

$$\left[\frac{m(n-m)}{n}\right]^{-1/2} = [np(1-p)]^{-1/2}(1+R)$$

where the magnitude of R is that of y/n, which in turn is of the magnitude of $1/n^{1/2}$ [see (2.40)]. Therefore,

$$\frac{\exp[c^*(n)/n]}{K[m(n-m)/n]^{1/2}} = \frac{1}{K[np(1-p)]^{1/2}} + R^*$$

where R^* goes to zero as fast as $1/n$. Finally, if we write

$$\left(1+\frac{y}{np}\right)^{-m}\left[1-\frac{y}{n(1-p)}\right]^{-(n-m)}$$

$$= \exp\left\{-m\log\left(1+\frac{y}{np}\right) - (n-m)\log\left[1-\frac{y}{n(1-p)}\right]\right\}$$

then, with the first two terms of the Taylor expansion of $\log(1+x)$, we get

$$\left(1+\frac{y}{np}\right)^{-m}\left[1-\frac{y}{n(1-p)}\right]^{-(n-m)} = e^{-(1/2)x_m^2} + R^{**}$$

where R^{**} goes to zero at the speed of $1/n^{1/2}$. Upon collecting all terms, we get, that for all m satisfying (2.38) [or (2.40)],

$$(2.42) \qquad P(k_A(n)=m) = \frac{1}{K[np(1-p)]^{1/2}}e^{-(1/2)x_m^2} + \frac{c}{n}$$

where c is bounded in n. Since the requirement made after (2.39) about the error term is satisfied, we can neglect the error terms when utilizing (2.42) in (2.37). Hence, as $n \to +\infty$, the limit of the probability in (2.37) is equal to

$$(2.43) \qquad \lim \sum{}' \frac{1}{K[np(1-p)]^{1/2}}e^{-(1/2)x_m^2}$$

Note that by the definition of x_m in (2.40),

$$x_{m+1} - x_m = [np(1-p)]^{-1/2}$$

and thus (2.43) can also be written as

$$(2.44) \qquad \lim \sum{}' \frac{1}{K}e^{-(1/2)x_m^2}(x_{m+1} - x_m)$$

where Σ', defined originally as summation over m satisfying (2.38) or (2.40), gets the meaning of summation over the x_j such that $a < x_1 < x_2 < \cdots < x_m < \cdots < b$, and $n \to +\infty$ is equivalent to $x_{m+1} - x_m \to 0$ for all m. Therefore, the sum in (2.44) is a Riemann sum and the limit there is the corresponding integral (of the function $(1/K) \exp[-(1/2)x^2]$). In other words, we established that, as $n \to +\infty$, the limit of the probability in (2.37) is

$$\frac{1}{K} \int_a^b e^{-(1/2)x^2} \, dx$$

It is left to the reader as Exercise 38 to show that $K = (2\pi)^{1/2}$ [the reason for which is that for large b and with $a = -b$, the Chebyshev inequality implies that the limit of the probability in (2.37) should be close to 1; on the other hand, we have seen in Section 2.7 that the integral above approaches 1 as $b = -a \to +\infty$ when $K = (2\pi)^{1/2}$].

Arguing with the Chebyshev inequality, one gets that the result remains valid if $a = -\infty$ or $b = +\infty$, whose details are omitted.

The theorem is established.   ▲

2.10   THE POISSON APPROXIMATION TO THE BINOMIAL PROBABILITIES

We remarked that the normal approximation to the binomial probabilities is applicable whenever $np(1-p) \geq 6$. In the present section we develop another approximation in which the emphasis is on p to be "small." This approximation, in contrast to the normal approximation, is usually applied to the individual terms $P(k_A(n) = m)$. The actual result is as follows.

Theorem 2.12   If $n \to +\infty$ and $p \to 0$ in such a way that $np \to \lambda > 0$, then for $m = 0, 1, 2, \ldots$,

$$P(k_A(n) = m) = \binom{n}{m} p^m (1-p)^{n-m} \to \frac{\lambda^m e^{-\lambda}}{m!}$$

Remark   The terms on the right-hand side in the limit above are called the *Poisson probabilities* (after a French mathematician). The Poisson approximation is accurate for most practical purposes when $np \leq 5$ and $n \geq 20$.

Proof of the theorem: If we write $np = \lambda + \varepsilon_n$, then, by assumption, $\varepsilon_n \to 0$ as $n \to +\infty$. Consequently, $(np)^m \to \lambda^m$ for every fixed m. Hence, since

$$\binom{n}{m}p^m(1-p)^{n-m} = \frac{n(n-1)\cdots(n-m+1)}{m!}p^m(1-p)^{n-m}$$

$$= \frac{(1-1/n)(1-2/n)\cdots[1-(m-1)/n]}{m!}(np)^m\left(1-\frac{\lambda+\varepsilon_n}{n}\right)^{n-m}$$

we get, as $n \to +\infty$,

$$\lim\binom{n}{m}p^m(1-p)^{n-m} = \frac{\lambda^m}{m!}\lim\left(1-\frac{\lambda+\varepsilon_n}{n}\right)^{n-m}$$

Finally, we know from calculus that, as $n \to +\infty$,

$$\lim\left(1-\frac{\lambda+\varepsilon_n}{n}\right)^n = e^{-\lambda}$$

which limit relation remains valid if we change the exponent n to $n-m$ with m fixed. This establishes the desired limiting form and completes the proof. ▲

Example 2.22 Assume that the probability that someone buys a specific expensive item is 0.003 if the person is aware that it is available. The company producing this item sends out 1000 pamphlets to randomly chosen addresses. What is the probability that people from at least two of these addresses order the item?

If A is the event that a person knowing about this item buys it, then $P(A) = 0.003$. In the example, the experiment (of informing prospective customers) concerning A is repeated $n = 1000$ times, independently of each other (because the population is large, there is practically no difference between selecting the addresses at random "with or without replacement"). With previous notations, we sought $P(k_A(1000) \geq 2)$. By turning to the complement,

$$P(k_A(1000) \geq 2) = 1 - P(k_A(1000) = 0) - P(k_A(1000) = 1)$$

Now, since $p = 0.003$, $n = 1000$, and thus $\lambda = np = 3$, Theorem 2.12 is applicable, which yields

$$P(k_A(1000) \geq 2) = 1 - e^{-3} - 3e^{-3} = 0.801 \qquad ▲$$

Example 2.23 Assume that, on average, you receive 3.6 pieces of mail a day. Is it justified to complain at the post office if the mail was not delivered? What if you received no mail (i) on two consecutive days or (ii) twice within a 3-month period?

It is remarkable that with so little information one can start and actually solve this problem. One can view correspondence as follows. Each day people and businesses decide with some (undetermined) probability p

that they will write to you. There are n (again undetermined) prospective correspondents, who act independently of each other. Therefore, the number k of pieces of mail received on a given day is the frequency of n independent repetitions of the specific event that someone writes to you. Because n is large, $k/n \sim p$ (Chebyshev's inequality), that is, $k \sim np$. Experience shows that k is 3.6 on average, so $\lambda = np = 3.6$. Hence, by the Poisson approximation,

$$P(\text{no mail}) = P(k = 0) \sim e^{-3.6} = 0.027$$

This means that, in every 100 days, you can expect no mail on 2 or 3 days. So you cannot complain if once in a while you have no mail; in fact, the situation in (ii) is exactly what you can anticipate, so you cannot complain in that case either. However, by assuming that mail received on one day has no influence on the amount of mail received the next day (or any other day),

$$P(\text{no mail on two consecutive days}) = P^2(k = 0) \sim 0.027^2 = 7.5 \times 10^{-4}$$

which can occur as rarely as once in about a 4-year period.   ▲

The interesting fact in the preceding example is that neither n nor p was known; therefore, the exact binomial formula could not be applied.

2.11 EXERCISES

1. $P(A) = 0.2$, $P(B) = 0.5$, and $P(A \cup B) = 0.6$. Find $P(A \mid B)$.

2. A pair of fair dice is rolled. Find the probability that at least one shows a 5, given that the sum of the two faces is 10.

3. Two (distinct) numbers are selected at random from the numbers 1 through 20. Find the probability that they are both odd numbers, given that their sum is even.

4. Pick a point x at random on the interval $(0, 1)$. Find $P(x < 1/2 \mid x > 1/4)$.

5. The probability that the stock of company A hits a new record on the stock market in a particular period is 0.3, whereas for company B this probability is 0.6. Given that the stock of company B hit a new record, a new record for company A is of probability 0.4. Find the probability that company B will have a new record for its stock in the period in question, given that the stock of company A reached a new record.

6. An urn contains six red and four white balls. If three balls are selected from the urn at random, one by one and without replacement, what is the probability that the first one is white and the other two are red?

7. From a group of five boys and three girls, four are selected at random, one after another, and then seated in a row in the order of selection. What is the probability that boys and girls alternate in the four seats?

8. Show that for arbitrary events A_1, A_2, and A_3, the events

(i) A_1, $A_2 \cap A_1^c$, and $A_1^c \cap A_2^c$

(ii) A_1, $A_2 \cap A_1^c$, $A_3 \cap A_1^c \cap A_2^c$, and $A_1^c \cap A_2^c \cap A_3^c$

form complete systems (partitions). Generalize this method for constructing complete systems from n arbitrary events.

9. Urn 1 contains three red and five white balls, while urn 2 is composed of five red and two white balls. One urn is picked at random, and one ball is chosen from this urn. Find the probability that a white ball is chosen.

10. Once again, urn 1 contains three red and five white balls, and urn 2 is composed of five red and two white balls. A person picks one ball at random from urn 1 and, without looking at its color, places it into urn 2. One ball is then picked at random from urn 2. What is the probability that this second choice results in a red ball?

11. A ball is picked at random from an urn containing one red and two white balls. If a red ball is chosen, a number is picked at random from the integers 1 through 10; otherwise a number is selected at random from the integers 6 through 10. Find the probability that the number (i) exceeds 7, and (ii) is even.

12. If in the selection procedure of Exercise 9 the ball chosen is white, what is the probability that it was chosen from urn 1?

13. Pick up at random one of two dice. One is regular, but on the other one two faces are marked by each of the numbers 1, 2, and 6. If the die chosen is rolled and the number 6 comes up, what is the probability that it is the regular die?

14. From an urn containing five red and three white balls one is picked at random and the ball, together with another one of the same color, is placed back into the urn. A second ball is then picked at random from the urn. Find the probability that the first choice resulted in a white ball, given that the second is red.

15. Assume that a disease, for whose detection a simple test is available, has spread to 30% of the population. The test is 98% effective in detecting the disease when present, and the test shows the disease only for 1% of those persons who do not suffer from it. Find the probability that someone is free from the disease in question, given that the test is positive for the patient. Compare the conclusion with that in Example 2.21 and explain the difference.

16. Show that A and B are independent if $P(A) = 0.6$, $P(B) = 0.5$, and $P(A \cup B) = 0.8$.

17. Let A, B, and C be independent events with $P(A) = 0.2$, $P(B) = 0.6$, and $P(C) = 0.7$. Evaluate $P(A \cup B \cup C)$.

18. Pick a number x from the integers 1 through 30. Let A be the event that x is divisible by 3, and C that x is divisible by 7. Show that A and C are dependent. What about the independence of A and D, where D is the event that x is divisible by 8?

19. Pick a number x from the integers 1 through 300. Find the probability that x is an even number which is not divisible by 3.

20. Pick a number x at random from the integers 1 through 200. Let A be the event

that x is between 76 and 175, B that x is between 1 and 100, and C that x is between 51 and 150 (in all cases, "between a and b" means that both a and b are included). Show that (i) B and C are independent, (ii) A is independent of both $B \cap C$ and $B \cup C$, but (iii) both A and B, and A and C are dependent.

21. Let $P(A) = P(B) = P(C) = 1/2$, $P(A \cap B \cap C) = (1/2)P(B \cap C) = (1/2)$ $P(A \cap C) = (1/2)P(A \cap B)$ and $P(A \cap (B \cup C)) = (1/2)P(B \cup C)$. Show that A, B, and C are independent. What is the role of 1/2 in the conditions?

22. Teams A and B play a best-of-seven series. If the probability that A wins against B in any one game is 0.6, and if the outcomes of different games are independent of each other, what is the probability that A wins the series? What is the probability that the series ends in five games?

23. Player A bets against the house with unlimited resources, and A wins \$1 if a fair coin lands on tails; otherwise he loses \$1. Assume A initially had \$3 only. What is the probability that A can play more than seven games without requiring credit?

24. Roll two regular dice five times. What is the probability that exactly three times the sum of the two faces up is 7?

25. In a multiple choice test each question can be answered in three different ways. If a student is completely unprepared for the test, what is the probability that he will give correct answers to at least 10 of 25 questions?

26. In a test consisting of 50 questions student S gives a correct answer to each question with probability 1/4, and whether or not his answers are correct to different questions are independent. One fails the course if fewer than 20 answers are correct. Use the normal approximation to the binomial probabilities to evaluate the probability that S fails the test.

27. Let k be the frequency of heads in 60 tosses of a fair coin. Give an exact formula for $P(26 \leq k \leq 34)$, and approximate its value by the normal distribution function.

28. A university wants to admit 600 students for its first-year class. Knowing from past experience that on the average, 60% of those accepted for admission will actually attend, the university approves the application of 1000 students. Use the normal approximation to the binomial probabilities to compute the probability that the actual size of the first-year class will not exceed 600. (Hint: View the experience of the university as if every student whose application is approved would decide with probability 0.6, and independently of one another, whether to attend that particular university.)

29. Evaluate d in the approximation

$$\sum_{k=0}^{170} \binom{400}{k} \sim 2^{400}d$$

30. An opinion poll is conducted in which, owing to the large size of the population, it is assumed that individuals questioned give a positive answer to a particular question with the same probability and independently of each other. How many individuals should be interviewed in order to be assured with

probability 0.95 that the relative frequency of those in favor of candidate A should be closer to the true proportion p of those in the whole population than 0.03, assuming that $0.3 < p < 0.7$?

31. Assume one wants to test whether a coin is fair or not by the following rule: Toss the coin 100 times; if the frequency of heads is between 42 and 58 (both inclusive), the coin is accepted as fair, otherwise termed biased. What is the probability that a coin is accepted as fair if it lands heads with probability 0.55?

32. A manufacturer finds that, on the average, 2% of his products are defective. What is the probability that out of 1000 products, selected at random from a very large lot of this manufacturer, 15 or more are defective?

33. Assume the probability of a man of age 40 dying within a year is 0.01. If an insurance company has 50,000 policies on men of age 40, approximate the probability by the normal distribution function that the number of claims to the company from beneficiaries of men of age 40 within a year will be beween 400 and 600.

34. Assume the probability of a man of age 40 dying within a year is 0.01. If someone knows 50 men of age 40, what is the probability that at least one of these acquaintances will die within a year? Interpret the result, and compare its meaning with Exercise 33.

35. A large metropolitan area is patroled by n policemen, who circulate at random and independently of each other in the whole area in such a manner that every part of the district is visited by any one of the policemen with the same probability p. The population observed that on the average, one policeman is in their neighborhood in every half hour. Why is it justified to use Poisson probabilities to compute the probability that exactly k policemen visit a particular neighborhood in a half-hour period? Use these probabilities to compute the probability that (i) no policeman, (ii) exactly one, and (iii) exactly two policemen turn up in a particular neighborhood in a given half-hour period.

36. Use the Poisson approximation to the binomial probabilities for computing

$$\binom{1000}{3}(0.002)^3(0.998)^{997}$$

37. Compare the normal and the Poisson approximations to the sum

$$\sum_{k=3}^{8}\binom{100}{k}(0.05)^k(0.95)^{100-k}$$

38. Complete the argument at the end of Section 2.9 for establishing that $K = (2\pi)^{1/2}$.

3
Random Variables

3.1 THE CONCEPT OF RANDOM VARIABLES

In many instances we are interested not only in qualities, but also in quantitative measurements of randomly selected items or individuals. For example, on the same questionnaire of an opinion poll people might be asked their political views (qualitative questions, which can be answered yes or no) as well as their income, number of years of education, number of children, and so on. Similarly, in addition to asking whether a randomly selected car is good or defective, a manufacturer (when setting up the terms of a warranty) would like to know the cost of repairs required in the next 12 months, say. When the answer to a question posed in connection with a random experiment is given by a number, we speak of random variables (random, because it is related to a random experiment, and it is a variable, because the answer will vary with the individual members of the experiment). When one looks at this vague definition more closely, one recognizes that we, in fact, speak of a function defined on Ω; that is, to every outcome of the experiment, we assigned a value. For purely mathematical reason, we have to add additional restriction to functions on Ω to become random variables. This is expressed in the following definition.

Definition 3.1 A function $X = X(\omega)$, defined on the set $\Omega = \{\omega\}$ of all possible outcomes of an experiment, is a *random variable* if for every real number z, the subset

$$(3.1) \qquad \{X \leq z\} = \{\omega: \omega \in \Omega, X(\omega) \leq z\}$$

of Ω is a member of the collection $\mathcal{A}$ of events.

In most cases we denote random variables using capital letters from the last letters of the alphabet. We usually suppress the variable ω, and when it does not lead to confusion, sets are abbreviated as on the left-hand side of (3.1).

In the present book we do not bother much about the latter part of Definition 3.1. Rather, we apply Definition 3.1 in the form of saying that a random variable X is a function on Ω, and, for every real number z,

(3.2) $$F(z) = P(X \le z)$$

is defined.

Definition 3.2   The function $F(z)$ of (3.2) (which is always defined for all real z) is called the *distribution function* of the random variable X.

The next section is devoted to the study of distribution functions. We close this section with a few examples.

First notice that we have encountered random variables and distribution functions in the preceding chapters. When the experiment was to select n items from a lot (with or without replacement), we denoted by X the number of type I items obtained. That is, to every outcome (selection of n) we assigned a number (X); in other words, we investigated a random variable (we shall return to this case in Section 3.4). Also, if n repetitions of the same experiment are considered as a single experiment, then the frequency $k_A(n)$ of the event A is a random variable whose distribution function is approximated in (2.26) (the case of $a = -\infty$) by the function $N(x)$, which at that time, without further explanation, we called the *standard normal distribution function*. More is said about this in Section 3.6.

Additional examples are given below.

Example 3.1   The cost of producing an electric light bulb is 40 cents, and a bulb is claimed to last 1000 (burning) hours. A full refund is promised if the bulb does not last longer than 600 hours. What is the profit on a bulb if it is sold for 60 cents?

The profit X is a random variable because for every bulb it takes on two possible (random) values: If a bulb can be used longer than 600 hours, then $X = 20$ cents (price minus cost of production); otherwise, $X = -40$ cents (the cost is lost through the refund). This can also be expressed through the (random) life Y of the bulb:

$$X = \begin{cases} 20 & \text{if } Y > 600 \\ -40 & \text{if } Y \le 600 \end{cases}$$

Therefore, $P(X = -40) = P(Y \leq 600)$ and $P(X = 20) = P(Y > 600) = 1 - P(Y \leq 600)$. Consequently, by the single value $F(600)$ of the distribution function $F(z) = P(Y \leq z)$, the profit X can be "predicted" (by the methods of Section 2.6). ▲

Example 3.2   The size of a bacterium is measured under a microscope, and it is recorded to be 3.86 units. What can we say about the accuracy of this result?

The recorded value 3.86 consists of the "true" size s plus a measurement error X, which is random since this contribution to the measured size varies from time to time (and from individual to individual). It is due to the inaccuracy of the instrument as well as to the experience (or inexperience) of the person measuring the size. Hence $s + X$ is found to be 3.86 rather than s. Now, if based on this measurement we would like to know whether $s \geq 3.8$, say, we actually ask whether $X \geq 0.06$. The latter question, however, can only be answered in probabilistic terms, that is, by $P(X \geq 0.06)$. In other words, questions concerning s transform into questions regarding X to which answers can be given in terms of the distribution function $F(z) = P(X \geq z)$. [It is remarkable that one can accurately determine the distribution function $F(z)$; see Section 5.4.] ▲

3.2   PROPERTIES OF DISTRIBUTION FUNCTIONS

We defined the distribution function $F(z)$ of a random variable X at (3.2). Its basic properties are given in the following three theorems.

Theorem 3.1   If the distribution function of the random variable X is $F(z)$, then, for all $a < b$,

$$P(a < X \leq b) = F(b) - F(a)$$

Theorem 3.2   Every distribution function $F(z)$ satisfies the following properties:
(i) $0 \leq F(z) \leq 1$.
(ii) $F(z)$ is nondecreasing.
(iii) $\lim_{z = +\infty} F(z) = 1$ and $\lim_{z = -\infty} F(z) = 0$.
(iv) $F(z)$ is continuous from the right $[F(z) = F(z + 0)$ for all $z]$.

Theorem 3.3   With the notations of Theorem 3.1, for every fixed number b,

$$P(X = b) = F(b) - F(b - 0)$$

where $F(b - 0)$ is the left-hand limit of $F(z)$ at b. In particular, if $F(z)$ is continuous at $z = b$, then $P(X = b) = 0$.

Before proving these theorems, let us look at some examples for their illustration.

Example 3.3   Suppose that the distribution function of X

$$F(z) = \begin{cases} 0 & \text{if } z < 1 \\ 1 - \dfrac{1}{z^2} & \text{if } z \geq 1 \end{cases}$$

Let us determine $P(2 < X \leq 3)$ and $P(2 < X < 3)$.

By Theorem 3.1, $P(2 < X \leq 3) = F(3) - F(2) = (1 - 1/9) - (1 - 1/4) = 5/36$. Now, $P(2 < X \leq 3) = P(2 < X < 3) + P(X = 3)$, and thus the difference between the two desired probabilities is $P(X = 3)$. However, since $F(z)$ is continuous at $z = 3$ (at any point actually), $P(X = 3) = 0$ from Theorem 3.3. Therefore, both answers are 5/36.   ▲

Example 3.4   Let the distribution function of X be

$$F(z) = \begin{cases} 0 & \text{if } z < 0 \\ \dfrac{1}{2}z & \text{if } 0 \leq z < 1 \\ 1 - \dfrac{1}{(2z)^2} & \text{if } z \geq 1 \end{cases}$$

Let us evaluate $P(X = 0)$, $P(X = 1)$, and $P(X \geq 2)$.

For evaluating the first two of the probabilities in question, we apply Theorem 3.3. Since $F(z)$ is continuous at $z = 0$, $P(X = 0) = 0$. On the other hand, $F(z)$ is discontinuous at $z = 1$, and thus

$$P(X = 1) = F(1) - F(1 - 0) = \left(1 - \frac{1}{4}\right) - \lim_{z=1}\left(\frac{1}{2}z\right) = \frac{1}{4}$$

Finally, by turning to the complement,

$$P(X \geq 2) = 1 - P(X < 2) = 1 - [P(X \leq 2) - P(X = 2)]$$

which equals $1 - F(2) = 1/16$, because, by the continuity of $F(z)$ at $z = 2$, $P(X = 2) = 0$.   ▲

Example 3.5   The distribution function of X is $F(z) = z$ for $0 < z < 1$. Is $F(z)$ uniquely determined for all z's?

The answer is yes. Since $F(0) = 0$ and $F(1) = 1$, by parts (i) and (ii) of Theorem 3.2, $F(z) = 0$ for all $z \leq 0$, and $F(z) = 1$ for all $z \geq 1$.   ▲

Example 3.6   The functions
(i) $F_1(z) = \sin z$ for $0 \leq z \leq \pi$,

(ii) $F_2(z) = z^2$ for $0 \leq z \leq 2$, and
(iii) $F_3(z) = z - 1$ for $1/2 < z \leq 2$,
cannot be parts of distribution functions. Why?

All of them violate one rule or another in Theorem 3.2. That is, $F_1(z)$ is decreasing for $(1/2)\pi < z \leq \pi$, $F_2(z) > 1$ if $z > 1$, and $F_3(z) < 0$ for $1/2 < z < 1$.   ▲

Example 3.7   For which values of the constants A and B is the function $F(z) = A + B \arctan z$ a distribution function?

Since

$$\lim_{z = -\infty} \arctan z = -\frac{1}{2}\pi \qquad \text{and} \qquad \lim_{z = +\infty} \arctan z = \frac{1}{2}\pi$$

the function $(1/\pi)[\arctan z + (1/2)\pi]$ satisfies properties (i) to (iv) of Theorem 3.2. Hence $F(z)$ can be a distribution function only with the constants $A = 1/2$ and $B = 1/\pi$. The fact that this is indeed a distribution function will be seen in Section 3.9.   ▲

We now turn to the proofs of Theorems 3.1 to 3.3.

Proof of Theorem 3.1:   Define the events A and B as $A = \{X \leq a\}$ and $B = \{X \leq b\}$. Then $A \subset B$, and $\{a < X \leq b\} = A^c \cap B$. Hence, by Theorem 1.3,

$$P(a < X \leq b) = P(B) - P(A) = F(b) - F(a)$$

which was to be proved.   ▲

For a substantial part of the rest of the proofs we need the following result.

Lemma   If the events A_1, A_2, … satisfy
(i) $A_1 \subset A_2 \subset \cdots$, then $\lim_{n = +\infty} P(A_n) = P(\bigcup_{j=1}^{+\infty} A_j)$
(ii) $A_1 \supset A_2 \supset \cdots$, then $\lim_{n = +\infty} P(A_n) = P(\bigcap_{j=1}^{+\infty} A_j)$

Proof: If $A_1 \subset A_2 \subset \cdots$, then

$$(3.3) \quad \bigcup_{j=1}^{+\infty} A_j = A_1 \cup (A_1^c \cap A_2) \cup (A_2^c \cap A_3) \cup \cdots \cup (A_{k-1}^c \cap A_k) \cup \cdots$$

where the terms of the union on the right-hand side are mutually exclusive. Hence, by axiom $(A3)$ of probability,

$$(3.4) \quad P\left(\bigcup_{j=1}^{+\infty} A_j\right) = \sum_{k=1}^{+\infty} P(A_{k-1}^c \cap A_k) = \lim_{n = +\infty} \sum_{k=1}^{n} P(A_{k-1}^c \cap A_k)$$

where, for uniformity, we wrote $A_1 = A_0^c \cap A_1$ with $A_0^c = \Omega$. But for the same reason as at (3.3),

$$\bigcup_{k=1}^{n} (A_{k-1}^c \cap A_k) = \bigcup_{j=1}^{n} A_j = A_n$$

and thus, by axiom (A3),

$$P(A_n) = \sum_{k=1}^{n} P(A_{k-1}^c \cap A_k)$$

Part (i) of the lemma now follows from (3.4). On the other hand, part (ii) follows from part (i) by simply observing that if $A_1 \supset A_2 \supset \cdots$, then the complements A_j^c satisfy $A_1^c \subset A_2^c \subset \cdots$, and, by De Morgan's law (Section 1.2), $\bigcup A_j^c = (\bigcap A_j)^c$. Therefore, by part (i) of the lemma,

$$\lim_{n=+\infty} P(A_n^c) = P\left(\bigcup_{j=1}^{+\infty} A_j^c\right) = P\left(\left(\bigcap_{j=1}^{+\infty} A_j\right)^c\right) = 1 - P\left(\bigcap_{j=1}^{+\infty} A_j\right)$$

which is the conclusion in part (ii). This completes the proof.   ▲

Proof of Theorem 3.2:   Because $F(z)$ is a probability, (i) is immediate from axiom (A1).

In part (ii), we have to prove that, if $a < b$, then $F(a) \le F(b)$. For proving this, again define $A = \{X \le a\}$ and $B = \{X \le b\}$. Then $A \subset B$, and thus, in view of Theorem 1.3, $P(A) \le P(B)$. But $P(A) = F(a)$ and $P(B) = F(b)$, which completes the proof of part (ii).

In the proofs of parts (iii) and (iv), we apply the lemma.

First let $A_n = \{X \le n\}, n = 1, 2, \dots$. Then $A_1 \subset A_2 \subset \cdots$, and the union of the A_j is Ω. Hence, by part (i) of the lemma,

$$\lim_{n=+\infty} F(n) = \lim_{n=+\infty} P(A_n) = P(\Omega) = 1$$

This implies that $F(z) \to 1$ as $z \to +\infty$ in an arbitrary manner, because we have just proved in part (ii) that $F(z)$ is nondecreasing, and thus the limit of $F(z)$ exists as $z \to +\infty$.

Next, if $A_n = \{X \le -n\}$, $n = 1, 2, \dots$, then $A_1 \supset A_2 \supset \cdots$, and the intersection of the A_j is empty. Consequently, part (ii) of the lemma yields

$$\lim_{n=+\infty} F(-n) = \lim_{n=+\infty} P(A_n) = P(\emptyset) = 0$$

Just as above, it now follows that $F(z) \to 0$ as $z \to -\infty$ from the monotonicity of $F(z)$.

Finally, let $A_n = \{b < X \le b + 1/n\}$, where b is a fixed value. Then $A_1 \supset A_2 \supset \cdots$, the intersection of the A_j is empty, and by Theorem 3.1, $P(A_n) = F(b + 1/n) - F(b)$. Hence, by part (ii) of the lemma,

$$\lim_{n=+\infty} F\left(b+\frac{1}{n}\right) - F(b) = \lim_{n=+\infty} P(A_n) = P(\varnothing) = 0$$

that is, the limit of $F(z)$ as z approaches b from the right is $F(b)$ [by the monotonicity of $F(z)$, it again suffices to approach b on a specific sequence]. This completes the proof.   ▲

Proof of Theorem 3.3:   This proof is an easy consequence of the lemma. Namely, if we define $A_n = \{b - 1/n < X \le b\}$, then $A_1 \supset A_2 \supset \cdots$, the intersection of the A_n is the set $\{X = b\}$, and by Theorem 3.1, $P(A_n) = F(b) - F(b - 1/n)$. Hence, by part (ii) of the lemma,

$$P(X = b) = \lim_{n=+\infty} P(A_n) = F(b) - \lim_{n=+\infty} F\left(b - \frac{1}{n}\right)$$

where the last limit is indeed $F(b - 0)$, from the monotonicity of $F(z)$. The proof is completed.   ▲

3.3   THE EXPONENTIAL DISTRIBUTION AS LIFE DISTRIBU-TION

In all examples of the preceding section, the distribution function of X was "made up." In the present section we determine the distribution function of X for an important practical model.

Let time be measured from the point on when person A purchases an accident insurance policy. Let X be the random time up to the death of A from an accident. We are to prove that the distribution function of X,

(3.5) $F(z) = P(X \le z) = 1 - e^{-az} \qquad z \ge 0$

where $a > 0$ is an undetermined real number and $F(z) = 0$ for $z < 0$. The distribution function $F(z)$ just introduced is called the *exponential distribution function*.

What is the mathematical translation of death by an accident? This can easily be expressed in terms of conditional probabilities. That is, accident, as opposed to aging and illness, means that passage of time has no influence on its occurrence. In other words, if A has lived t time units, A's chance of surviving another s time units is the same as if A had just purchased insurance (and thus time would start at this point). In other words,

(3.6) $P(X > t + s \mid X > t) = P(X > s) \qquad t, s \ge 0$

By the definition of conditional probabilities, this can also be written as

$$\frac{P(X > t + s \text{ and } X > t)}{P(X > t)} = P(X > s)$$

in which the numerator simplifies to $P(X > t + s)$, because in view of $t, s \geq 0$, $\{X > t + s\} \subset \{X > t\}$, and thus

(3.7) $\qquad P(X > t + s) = P(X > s)P(X > t) \qquad t, s \geq 0$

All of the events occurring in (3.7) are complements of events defining the distribution function; consequently, another form of (3.7) is

$$1 - F(t + s) = [1 - F(s)][1 - F(t)]$$

or, upon setting $G(z) = 1 - F(z)$,

(3.8) $\qquad G(t + s) = G(s)G(t) \qquad t, s \geq 0$

The claimed exponentiality of $F(z)$ now follows from the following result.

Lemma   Let $G(z)$ be a monotonic function satisfying (3.8). Then either $G(z)$ is identically zero for $z \geq 0$, or there is a number b such that $G(z) = e^{bz}$, $z \geq 0$.

Proof: If we replace t by $t_1 + t_2$, (3.8) yields

$$G(t_1 + t_2 + s) = G(s)G(t_1)G(t_2)$$

and, by induction, for every n and all $t_j \geq 0$, $1 \leq j \leq n$,

(3.9) $\qquad G(t_1 + t_2 + \cdots + t_n) = G(t_1)G(t_2)\cdots G(t_n)$

Choose each $t_j = 1$: we get $G(n) = G^n(1)$ for all $n \geq 1$. Next, choose each $t_j = 1/n$, yielding $G(1) = G^n(1/n)$ for arbitrary $n \geq 1$.

Note that it now follows that if $G(z)$ is not identically zero, $G(1) > 0$. First, $G(1)$ cannot be zero, because, by the assumed monotonicity of $G(z)$, from the two special cases given after (3.9), $G(1) = 0$ would imply $G(z) = 0$ for all $z \geq 0$. It is also impossible for $G(1)$ to be negative in view of $G(1) = G^2(1/2)$.

Because $G(1) > 0$, the two cases deduced from (3.9) can be combined into the form

(3.10) $\qquad G(z) = e^{bz}$

where $b = \log G(1)$, and z is either a positive integer or the reciprocal of a positive integer. However, continuing with (3.9), and choosing each $t_j = 1/m$, $n \neq m$, we get $G(n/m) = G^n(1/m)$, on the right-hand side of which (3.10) is applicable, which in turn transforms this last equation into the form in (3.10). In other words, (3.10) is valid for all positive rational numbers $z = n/m$.

Let now z be irrational. Then there are two sequences $z_{1,n}$ and $z_{2,n}$ of rational numbers such that $z_{1,n} < z < z_{2,n}$ for every n, and both $z_{1,n}$ and

$z_{2,n} \to z$ as $n \to +\infty$. Consequently, by the monotonicity of $G(z)$, either

$$\exp(bz_{1,n}) = G(z_{1,n}) \le G(z) \le G(z_{2,n}) = \exp(bz_{2,n})$$

or

$$\exp(bz_{2,n}) = G(z_{2,n}) \le G(z) \le G(z_{1,n}) = \exp(bz_{1,n})$$

for all n. Upon letting $n \to +\infty$, the two extreme sides tend to e^{bz}, and thus $G(z) = e^{bz}$ for all $z \ge 0$, rational or irrational. This completes the proof.   ▲

Because in our original problem leading to (3.8) $G(z) = 1 - F(z)$, which should be decreasing, and thus $b < 0$ in (3.10), (3.5) is immediate from the lemma.

Note that in the translation of "accident" into the equation (3.6) the emphasis is that aging is not a factor. Therefore, we get the same conclusion [i.e., the distribution function is (3.5)] whenever aging has no influence on the magnitude of a random variable X. For example, within a well-chosen warranty period, the time up to the first service (repair) has this property, as do time intervals between most natural disasters and many other events.

The significance of the result of this section is that a simple nonmathematical assumption (the no-aging property) leads to a unique type of distribution function.

3.4   DISCRETE RANDOM VARIABLES

One special class of random variables that we deal with in some detail is that of the discrete random variables.

Definition 3.3   The random variable X, and its distribution function $F(z)$, are called *discrete* if the set of possible values of X is either finite or denumerably infinite.

There are, therefore, two significant sequences associated with a discrete random variable: (i) its set $z_1, z_2, \ldots$ of possible values and (ii) the probabilities $p_j = P(X = z_j)$. From the fact that the z_j are the values, and the only values, of X, it follows that

$$(3.11) \qquad p_j > 0 \text{ for all } j \qquad \text{and} \qquad p_1 + p_2 + \cdots = 1$$

where "all j" and the sum in the second part extend over as many terms as many values X has.

Definition 3.4   A sequence $\{p_j\}$ of numbers satisfying (3.11) is called a (probability) *distribution*.

The distribution $\{p_j\}$ and the distribution function $F(z)$ of a discrete random variable X uniquely determine each other through the relation

$$F(z) = \sum_{z_j \leq z} p_j$$

Note the structure of $F(z)$ in this relation: $F(z)$ has a jump of magnitude p_t at $z = z_t$ (if the points z_j are not dense in the sense that there is a neighborhood of z_t such that no other z_j falls into it), and $F(z)$ does not change (it is constant) until a new z_j is reached by z. Expressed in another way, $P(a < X \leq b) = 0$ if $a < b$ are such that no z_j falls between them, and if at least one z_j satisfies $a < z_j \leq b$, then

$$P(a < X \leq b) = \sum P(X = z_j) = \sum p_j$$

where the summation is over all j such that $a < z_j \leq b$.

Example 3.8 Assume the following distribution function of X:

$$F(z) = \begin{cases} 0 & \text{if } z < 0 \\ \dfrac{1}{2} & \text{if } 0 \leq z < 1 \\ 1 & \text{if } z \geq 1 \end{cases}$$

Hence X is a discrete random variable with values $z_1 = 0$ and $z_2 = 1$, and with distribution $p_1 = p_2 = 1/2$.

Since $F(z)$ has jumps at $z = 0$ and $z = 1$, and it is constant before 0, between 0 and 1, and after 1, $F(z)$ is a discrete distribution function. The points $z = 0$ and 1 of the jumps are the possible values of X, and the magnitudes of the jumps (1/2 at both 0 and 1) give the distribution of X. ▲

Example 3.9 The distribution function

$$F(z) = \begin{cases} 0 & \text{if } z < 0 \\ \dfrac{1}{2}z + \dfrac{1}{4} & \text{if } 0 \leq z < 1 \\ 1 & \text{if } z \geq 1 \end{cases}$$

is not discrete.

Although $F(z)$ again has jumps at $z = 0$ and $z = 1$, and is continuous elsewhere, and thus, by Theorem 3.3, $P(X = 0) = 1/4$ and $P(X = 1) = 1/4$ (the magnitudes of jump both at $z = 0$ and 1 are 1/4), and $P(X = z) = 0$ for all other z, where X is a random variable whose distribution function is $F(z)$, it does not follow that the set of possible values of X is limited to 0 and 1. In fact, because $F(z)$ is not constant between 0 and 1, we get for $0 < a < b < 1$,

$$P(a < X \leq b) = F(b) - F(a) = \left(\frac{1}{2}b + \frac{1}{4}\right) - \left(\frac{1}{2}a + \frac{1}{4}\right) = \frac{1}{2}(b - a) > 0$$

and thus the set of possible values of X includes all intervals (a, b) with $0 < a < b < 1$. Hence X is not discrete.   ▲

Example 3.10   The random variable X is discrete with values $-2, 0, 1$, and 3. Let us determine the distribution of X if $P(X = -2) = 1/2$ and $P(X = 0) = P(X = 1) = 2P(X = 3)$.

Put $p_1 = P(X = -2), p_2 = P(X = 0), p_3 = P(X = 1)$, and $p_4 = P(X = 3)$. Then, by (3.11), $p_1 + p_2 + p_3 + p_4 = 1$, and we know that $p_1 = 1/2$, $p_2 = p_3 = 2p_4$. Hence, $5p_4 = 1/2$, i.e., $p_4 = 0.1$, from which $p_2 = p_3 = 0.2$. Thus the distribution of X is $(0.5, 0.2, 0.2, 0.1)$.   ▲

We have encountered discrete random variables in the previous chapters without calling them such. Let us restate some important results in terms of the newly introduced concepts.

The Hypergeometric Distribution

Let a lot consist of M type I and $T - M$ type II items. Let us select t items from the lot at random without replacement. Let X be the number of type I items among those selected.

Then X is a discrete random variable whose possible values are $0, 1, 2$, $\ldots, m^* = \min(M, t)$. The distribution of X, as determined in Theorem 1.7, is

$$p_m = P(X = m) = \frac{\dbinom{M}{m}\dbinom{T - M}{t - m}}{\dbinom{T}{t}} \qquad 0 \leq m \leq m^*$$

This sequence of probabilities is called the *hypergeometric distribution*. Note that it has three parameters: T, M, and t, all of which are positive integers and that both t and $M < T$.

The Binomial Distribution

Let A be an event in connection with a random experiment. Let $p = P(A)$. Let us repeat the experiment n times independently of each other. If $X = k_A(n)$ is the frequency of A in these repetitions (i.e., the number of times when A occurred), X is a discrete random variable with possible values $0, 1, 2, \ldots, n$. Its distribution was found in Theorem 2.7:

$$p_m = P(X = m) = \binom{n}{m} p^m (1 - p)^{n-m} \qquad 0 \le m \le n$$

This distribution is a *binomial distribution*. It has two parameters n and p, where n is a positive integer and $0 \le p \le 1$.

Although we also encountered the binomial distribution in connection with selecting items with replacement from a lot containing two types (Section 1.8), it was later established in Section 2.5 (see particularly the discussion following Definition 2.6) that such a selection model is exactly the model of independent repetitions of selecting just one item.

The Poisson Distribution

The distribution of the random variable X, whose values are the nonnegative integers, is called *Poisson* if, with some number $\lambda > 0$, which is its sole parameter,

$$p_m = P(X = m) = \frac{\lambda^m e^{-\lambda}}{m!} \qquad 0 \le m < +\infty$$

So far we came across the Poisson distribution only as an approximation to the binomial distribution (Theorem 2.12). It is therefore not evident that the sequence above is indeed a distribution, that is, that the sum of the terms is 1. This fact, however, follows from the Taylor formula of calculus for e^λ:

$$e^\lambda = 1 + \lambda + \frac{\lambda^2}{2!} + \cdots + \frac{\lambda^m}{m!} + \cdots$$

Namely,

$$\sum_{m=0}^{+\infty} p_m = \sum_{m=0}^{+\infty} \frac{\lambda^m e^{-\lambda}}{m!} = e^{-\lambda} \sum_{m=0}^{+\infty} \frac{\lambda^m}{m!} = e^{-\lambda} e^\lambda = 1$$

Later in the chapter we shall discuss a model in which the Poisson distribution reappears, this time as the exact distribution (i.e., not just an approximation).

The Geometric Distribution

In Example 1.31 (and in the discussion following it), we solved a special case of the following waiting-time problem.

Let A be an event in connection with a random experiment, and let $p = P(A)$. We shall repeat the experiment in independent trials as many times as needed for A to occur. Let X be the number of repetitions required in this procedure; that is, $\{X = m\}$ means that A occurred in the mth repetition but that it failed in each of the first $m - 1$ repetitions. The random variable X, whose possible values are the positive integers, is evidently discrete. Its distribution is given by

$$p_m = P(X = m) = p(1 - p)^{m - 1} \qquad 1 \leq m < +\infty$$

This distribution is called the *geometric distribution*. It has the single parameter $0 < p < 1$. The random variable X is also known as the (discrete) waiting time (variable), since X tells how long we have to wait for A to occur.

The proof of the form $p_m = p(1 - p)^{m - 1}$ is immediate from the concept of independent repetitions of an experiment. That is, if we define A_j as "A occurs at the jth repetition," then these events are independent, $P(A_j) = p$ for each j, and $\{X = m\} = A_1^c \cap A_2^c \cap \cdots \cap A_{m-1}^c \cap A_m$. Upon taking probabilities in this last equation, the formula claimed for $P(X = m)$ now follows.

Note that, in principle, the case of "A never occurring" is also possible. However, since, for every $m \geq 1$,

$$\{X = +\infty\} \subset \{X > m\} = A_1^c \cap A_2^c \cap \cdots \cap A_m^c$$

and thus

$$P(X = +\infty) \leq (1 - p)^m \to 0 \qquad \text{as } m \to +\infty$$

it follows that $P(X = +\infty) = 0$. Consequently, the sequence $\{p_m\}$, $1 \leq m < +\infty$, is a distribution, implying that

$$1 + (1 - p) + (1 - p)^2 + \cdots + (1 - p)^n + \cdots = \frac{1}{p}$$

a formula well known in calculus (and which we have proved in Section 1.10). What is interesting here is that we proved this formula by a purely probabilistic argument.

We conclude this list with the following simple random variable, which serves as a method of translating events into random variables.

The Indicator (or Bernoulli) Random Variable

To an event A we assign the random variable

$$I(A) = \begin{cases} 1 & \text{if } A \text{ occurs} \\ 0 & \text{if } A \text{ fails to occur} \end{cases}$$

Hence the distribution of $I(A)$ is $p_1 = P(I(A) = 1) = P(A)$ and $p_2 = P(I(A) = 0) = 1 - P(A)$. The random variable $I(A)$ is called the *indicator variable* (of A). It is also called the Bernoulli variable, for historical reasons.

Example 3.11   Each of 10 persons picks a "favorite" number (between 1 and 36 inclusive), and they go to different roulette tables to bet repeatedly on their own number. They set up a rule among themselves that the person who wins stops playing. What is the probability that exactly two of them will bet fewer than 20 times?

Because the 10 people play at different tables, we face 10 independent repetitions of the same experiment: repeated spins of the roulette wheel. Because they agreed on a common rule, each of them is interested in the occurrence of the same event A: that a specific number turns up (ultimately). Hence, if X is the number of persons among the 10 who will win in fewer than 20 bets, the distribution of X is binomial with parameters $n = 10$ and $p = P(B)$, where B is the event that a predetermined number will come up in fewer than 20 spins of the roulette wheel. This, however, can be computed with the help of a geometric distribution: the probability that fewer than 20 repetitions are required for an event to occur (a specific number to come up) whose probability is 1/38 (there are 36 numbers and two "zeros" on a roulette wheel). Hence

$$p = P(B) = 1 - \left(1 - \frac{1}{38}\right)^{19} = 0.40$$

and

$$P(X = 2) = \binom{10}{2} 0.4^2 \times 0.6^8 = 0.12 \qquad \blacktriangle$$

3.5   ABSOLUTELY CONTINUOUS DISTRIBUTION FUNCTIONS

Another class of distribution functions that we shall focus on is that of the absolutely continuous distribution functions.

Definition 3.5   Let $F(z)$ be the distribution function of X. Then $F(z)$, or X, is called *absolutely continuous* if $F(z)$ is continuous everywhere, and if, except perhaps at a finite number of points, $F(z)$ is differentiable. The function $f(z) = F'(z)$ (defined for all z except perhaps for a finite number of points) is called the *density function* of $F(z)$ or of X.

The exponential distribution function introduced in (3.5) is absolutely continuous with density function $f(z) = ae^{-az}$ if $z > 0$ and zero otherwise. Many other such functions will be discussed in detail in subsequent sections.

The basic properties of density functions are given in the following theorem.

Theorem 3.4   Every density function $f(z)$ satisfies the following conditions:
(i) $f(z) \geq 0$ (where defined)

(ii) $\int_{-\infty}^{+\infty} f(z)\, dz = 1$

(iii) $F(z) = \int_{-\infty}^{z} f(t)\, dt$

Proof: This theorem is an easy consequence of Theorem 3.2. Namely, since $f(z) = F'(z)$, and $F(z)$ is nondecreasing, we know from calculus that $f(z) \geq 0$. By one more appeal to calculus (called the fundamental theorem), we have

$$\int_{a}^{z} f(t)\, dt = F(z) - F(a)$$

Now, upon letting $a \to -\infty$, we get part (iii) of the theorem, from which part (ii) follows by letting $z \to +\infty$. This completes the proof.   ▲

Example 3.12   The distribution function

$$F(z) = \begin{cases} 1 - \dfrac{1}{2}e^{-2z} & \text{if } z \geq 0 \\ 0 & \text{otherwise} \end{cases}$$

is not absolutely continuous.

Even though $F(z)$ is differentiable at every z except at $z = 0$, $F(z)$ is not absolutely continuous because $F(z)$ is not continuous at $z = 0$.   ▲

Example 3.13   Suppose that $F(z)$ is an absolutely continuous distribution function with density function

$$f(z) = \begin{cases} \dfrac{3}{8}z^2 & \text{if } 0 < z < A \\ 0 & \text{otherwise} \end{cases}$$

Let us determine A.

Because $F(z)$ is a constant if $z < 0$ and also a constant if $z \geq A$ $[F'(z) = f(z) = 0]$, we must have $F(z) = 1$ if $z \geq A$ and $F(z) = 0$ if $z < 0$ [Theorem 3.2(iii)]. On the other hand,

$$F(z) = \int_{-\infty}^{z} f(t)\, dt = \frac{1}{8}z^3 \qquad \text{if } 0 \leq z < A$$

and thus the continuity of $F(z)$ at $z = A$ gives the equation $(1/8)A^3 = 1$, from which $A = 2$.   ▲

Example 3.14   The random variable X is absolutely continuous with density function

$$f(z) = \begin{cases} Az & \text{if } 0 < z < 3 \\ 0 & \text{otherwise} \end{cases}$$

Let us evaluate $P(1 < X < 2)$.

Because the distribution $F(z)$ of X is continuous, $P(1 < X < 2) = F(2) - F(1)$. On the other hand,

$$F(2) - F(1) = \int_1^2 f(z)\, dz = \frac{1}{2}Az^2 \Big]_1^2 = \frac{3A}{2}$$

Now, since the total integral of $f(z)$ is 1, and $f(z) = 0$ outside the interval $(0, 3)$, we have

$$1 = \int_0^3 Az\, dz = \frac{1}{2}Az^2 \Big]_0^3 = \frac{9A}{2} \quad \text{that is,} \quad A = \frac{2}{9}$$

Upon combining all results, we have $P(1 < X < 2) = 1/3$.   ▲

A new interpretation can be given to what we called geometric probabilities in Chapter 1. This is given below for the special case of choosing points from an interval.

The uniform distribution over an interval: Let X be a randomly chosen point of the interval (a, b). We say that X is uniformly distributed over (a, b) if it has a density function $f(z)$, which is constant over (a, b) and 0 outside it. By Theorem 3.4(ii) the constant must be $1/(b - a)$, and thus

$$f(z) = \begin{cases} \dfrac{1}{b-a} & \text{if } a < z < b \\ 0 & \text{otherwise} \end{cases}$$

Now, if $a < c < d < b$, then

$$P(c < X < d) = \int_c^d \frac{1}{b-a}\, dz = \frac{d-c}{b-a}$$

that is, the probability that X falls into the subinterval (c, d) is proportional to its length $d - c$, a property that we used as the definition of geometric probabilities.

The uniform distribution has a special role in probability theory, which is reflected in the following statement.

Theorem 3.5   Let X be a random variable with continuous distribution function $F(z)$. Then the new random variable $Y = F(X)$ is uniformly distributed over $(0, 1)$.

Proof: Since every distribution function is nondecreasing, the continuous distribution function $F(z)$ has an inverse $F^{-1}(z)$ [i.e., a function such that for all z, $F(F^{-1}(z)) = z$]. Therefore, the distribution function $F^*(z)$ of Y, for $0 < z < 1$, satisfies

$$F^*(z) = P(Y \le z) = P(F(X) \le z) = P(X \le F^{-1}(z)) = F(F^{-1}(z)) = z$$

On the other hand, in view of $Y = F(X)$, $0 \le Y \le 1$, and thus $F^*(z) = 0$ if $z < 0$ and $F^*(z) = 1$ if $z \ge 1$. Consequently, $F^*(z)$ is continuous everywhere, differentiable for all z except at $z = 0$ and 1, and $f^*(z)$, the derivative of $F^*(z)$, is constant (equals 1) over $(0, 1)$, and zero outside it; that is, $Y = F(X)$ is uniformly distributed over $(0, 1)$.   ▲

Theorem 3.5 can be utilized both for simplifying some mathematical arguments (general statements can be proved through the special case of uniform distributions) and for generating values of a random variable (i.e., random numbers) on a computer using a single program. That is, by generating values of a uniformly distributed random variable Y, through the equation $Y = F(X)$ [i.e., $X = F^{-1}(Y)$] one automatically gets values of X, which are distributed according to $F(z)$. Therefore, one can simulate on a computer any type of random behavior through a program for the uniform distribution.

3.6   THE NORMAL DISTRIBUTION

In Chapter 2 an approximation of the binomial distribution was developed in terms of the standard normal distribution function

$$(3.12) \qquad N(z) = (2\pi)^{-1/2} \int_{-\infty}^{z} e^{-t^2/2} \, dt$$

It was shown in Section 2.7 that $N(z)$ indeed satisfies the properties required for all distribution functions as given in Theorem 3.2. We also presented a table (Table 2.1) for the values of $N(z)$.

Now let X be a random variable with distribution function $N(z)$. Let m and $\sigma > 0$ be two numbers. We shall show that the distribution function of

$$(3.13) \qquad Y = \sigma X + m$$

is the function

$$(3.14) \qquad N(z;m,\sigma) = \frac{1}{\sqrt{2\pi}\,\sigma} \int_{-\infty}^{z} e^{-(t-m)^2/2\,\sigma^2} \, dt$$

The random variable Y is called a *normal* (random) *variable*, and the function $N(z;m,\sigma)$ is called the *normal distribution function* (with parameters m and $\sigma > 0$). The case $m = 0$ and $\sigma = 1$ reduces to the standard normal distribution.

Let us now establish the claim about the distribution function of Y. Since the distribution function of X is $N(z)$,

$$F(z) = P(Y \le z) = P(\sigma X + m \le z) = P\left(X \le \frac{z - m}{\sigma}\right)$$

$$= N\left(\frac{z - m}{\sigma}\right) = (2\pi)^{-1/2} \int_{-\infty}^{(z-m)/\sigma} e^{-t^2/2} \, dt$$

Let us substitute $y = \sigma t + m$. Then $dy = \sigma dt$ and the new limits of integration are $-\infty$ and $\sigma[(z - m)/\sigma] + m = z$; that is, the integral above becomes (3.14), as claimed.

Note that $N(z;m,\sigma)$ is everywhere differentiable. Hence a normal random variable has the density

(3.15) $$n(z;m,\sigma) = \frac{1}{\sqrt{2\pi}\,\sigma} e^{-(z - m)^2/2\sigma^2}$$

The normal distribution function is one of the most frequently recurring distribution function in probability theory and statistics. It is used as an approximation to a large variety of problems, not just to the binomial distribution (see Chapter 5), and it occurs in several contexts as the exact distribution function of a random variable. We shall, for example, outline an argument in Chapter 5, which shows that errors in laboratory measurements are normally distributed. Unfortunately, an exact proof requires mathematical tools far beyond the level of the current book.

The relation (3.13) is very significant in connection with using Table 2.1 in computations involving normal random variables.

Example 3.15 Let Y be a normal random variable with parameters $m = -2$ and $\sigma = 2$. Let us determine $P(0 < Y < 2)$.

In view of (3.13), $Y = 2X - 2$, where X is standard normal. Hence

$$P(0 < Y < 2) = P(1 < X < 2) = N(2) - N(1) = 0.1359$$

where the values $N(2)$ and $N(1)$ were taken from Table 2.1.   ▲

Example 3.16 Let Y be a normal random variable with parameters $m = -2$ and $\sigma = 2$. Let us show that $Z = 5 - 3Y$ is also normal, and let us evaluate $P(-1 < Z < 20)$.

Just as in Example 3.15, $Y = 2X - 2$, where X is standard normal. Consequently,

(3.16) $$Z = -6X + 11 = 6(-X) + 11$$

Now, if X is standard normal, so is $(-X)$ because of the symmetry of $n(z) = n(z;0,1)$ about zero (see also Theorem 2.11). Hence Z is normal with parameters $m = 11$ and $\sigma = 6$ [compare (3.16) with (3.13)], and thus

$$P(-1 < Z < 20) = P(-2 < (-X) < 1.5) = N(1.5) - N(-2)$$

$$= N(1.5) - [1 - N(2)] = 0.9104 \qquad \blacktriangle$$

We conclude this section with an example that is seemingly contradictory, yet it represents a quite appropriate application of normality. We shall "prove" in Chapter 5 that averages of random variables can be approximated very accurately by normal random variables. This means that for practical purposes, the distribution function of the average of random variables can be replaced by an appropriate normal distribution function. This is behind the assumption of the following example.

Example 3.17   Suppose that the average useful life Y of a Buick Apollo (car) is (at least asymptotically) normally distributed with parameters $m = 8$ and $\sigma = 1.6$, where Y is measured in years. Let us analyze the meaning of this assumption.

First, by (3.13), $Y = 1.6X + 8$, where X is a standard normal variable. Therefore, from the general property

$$P(8 + 1.6a < Y < 8 + 1.6b) = P(a < X < b) = N(b) - N(a)$$

and from Table 2.1 with varying a and b, we get, for example,

$$P(6.4 < Y < 9.6) = N(1) - N(-1) = 2N(1) - 1 = 0.6826$$
$$P(3.8 < Y) = 1 - N(-3) = N(3) = 0.9987$$
$$P(Y < 11.2) = N(2) = 0.9772$$

and so on. This means that if 100,000 Apollos were produced (the Apollo line was produced for a few years only around 1973), then about 68,260 were running for more than 6.4 years but less than 9.6 years, 99,870 were running for more than 3.8 years, and only about 3.280 can be expected to run for over 11.2 years.

The seemingly contradictory part of the assumption of the present example is that, in principle, $Y < 0$ is also possible. As a matter of fact, $P(Y < 0) = P(X < -5) = 1 - N(5)$, which, although not contained in Table 2.1, equals a number whose first five digits are zero, and thus it is small enough for 100,000 cars not to lead to something impractical. $\qquad \blacktriangle$

The careful reader can recognize that the ratio m/σ determines $P(Y < 0)$ for a normal variable. Hence, depending on m/σ, a normal variable Y can well describe a random quantity even if it is known to be positive.

3.7 MULTIVARIATE DISTRIBUTIONS

Several random variables X_1, X_2, ..., X_d can be viewed as a vector $(X_1, X_2, ..., X_d)$ with d components. The (multivariate, or d-variate, or joint) distribution function of the vector $(X_1, X_2, ..., X_d)$ is defined as

$$(3.17) \qquad F(z_1, z_2, ..., z_d) = P(X_1 \le z_1, X_2 \le z_2, ..., X_d \le z_d)$$

where each z_j runs through all real numbers.

Starting with (3.17), if one, or several, of the variables z_j, $1 \le j \le d$, are missing from $F(z_1, z_2, ..., z_d)$, we speak of lower-dimensional marginal distribution functions. More accurately, if for a fixed j, $1 \le j \le d$, $z_j \to +\infty$, $F(z_1, z_2, ..., z_d)$ converges to a function of $d - 1$ variables, which is actually the distribution function of the vector $(X_1, X_2, ..., X_{j-1}, X_{j+1}, ..., X_d)$. This can be deduced from the lemma of Section 3.2 in the same manner in which Theorems 3.2 and 3.3 were deduced from it; we therefore omit the details. Now letting another variable $z_t \to +\infty$, we get a $(d-2)$-dimensional distribution function, and so on. When all but one variables have been eliminated from $F(z_1, z_2, ..., z_d)$ by this limiting procedure, we get the (univariate marginal) distribution functions of the components X_k, $1 \le k \le d$.

Other properties, which can be proved in the same way as in the one-dimensional case, are as follows:

(i) Every (multivariate) distribution function is nondecreasing in each of its variables.

(ii) If all $z_j \to +\infty$, $1 \le j \le d$, then $\lim F(z_1, z_2, ..., z_d) = 1$.

(iii) If one, or several, $z_j \to -\infty$, then $\lim F(z_1, z_2, ..., z_d) = 0$.

If every component X_j of $(X_1, X_2, ..., X_d)$ is discrete, that is, the set of possible values of X_j is a sequence for every $1 \le j \le d$, we then speak of multivariate discrete vectors or distribution functions. Just as in the univariate case, it is more convenient to calculate probabilities concerning the discrete vector $(X_1, X_2 ..., X_d)$ by means of the (multivariate) distribution

$$(3.18) \quad p(m_1, m_2, ..., m_d) = P(X_1 = x_{1,m_1}, X_2 = x_{2,m_2}, ..., X_d = x_{d,m_d})$$

where x_{j,m_j} runs through all possible values of X_j, $1 \le j \le d$, than by means of the distribution function $F(z_1, z_2, ..., z_d)$. Of course, the two uniquely determine each other through the relation: The sum of those members $p(m_1, m_2, ..., m_d)$ of the distribution (3.18) for which $x_{j,m_j} \le z_j$, $1 \le j \le d$, equals $F(z_1, z_2, ..., z_d)$.

The basic properties of multivariate discrete distributions are two summation rules:

(i)   The sum of the terms in (3.18) over a single component m_j gives the $(d-1)$-dimensional distribution of the vector obtained from $(X_1, X_2, \ldots, X_d)$ by the removal of the component X_j.

(ii) The sum of all terms of (3.18) is 1.

Both of these rules are immediate from the axioms of probability.

Example 3.18   The random vector (X_1, X_2) is discrete, each component taking on the values 0, 1, and 2 only, with distribution $p(i, j) = P(X_1 = i, X_2 = j)$ as follows:

$$\begin{aligned}
p(0,0) &= 0.15, & p(0,1) &= 0.2, & p(0,2) &= 0.1 \\
p(1,0) &= 0.2, & p(1,1) &= 0.05, & p(1,2) &= 0.15 \\
p(2,0) &= 0.05, & p(2,1) &= 0.05, & p(2,2) &= 0.05
\end{aligned}$$

Let us determine the distribution of each of X_1 and X_2. Furthermore, let us evaluate $P(X_1 \le 1.8, X_2 \le 0.7)$, $P(1 \le X_1 < 2, 0.7 \le X_2 < 1.6)$, $P(X_1 + X_2 = 2)$, and $P(X_1 = 2 \,|\, X_2 \ge 1)$.

By property (i) of discrete distributions, row sums of the distribution $p(i, j)$ represent the distribution of X_1, and column sums give the distribution of X_2. Hence

$$\begin{aligned}
P(X_1 = 0) &= 0.45, & P(X_1 = 1) &= 0.4, & P(X_1 = 2) &= 0.15 \\
P(X_2 = 0) &= 0.4, & P(X_2 = 1) &= 0.3, & P(X_2 = 2) &= 0.3
\end{aligned}$$

Next, evidently,

$$P(X_1 \le 1.8, X_2 \le 0.7) = P(X_1 = 0, X_2 = 0) + P(X_1 = 1, X_2 = 0) = 0.35$$

$$P(1 \le X_1 < 2, 0.7 \le X_2 < 1.6) = P(X_1 = 1, X_2 = 1) = 0.05$$

$$\begin{aligned}
P(X_1 + X_2 = 2) = P(X_1 = 0, X_2 = 2) + P(X_1 = 1, X_2 = 1) + \\
P(X_1 = 2, X_2 = 0) = 0.2
\end{aligned}$$

and by the definition of conditional probabilities,

$$P(X_1 = 2 \,|\, X_2 \ge 1) = \frac{P(X_1 = 2, X_2 \ge 1)}{P(X_2 \ge 1)}$$

$$= \frac{P(X_1 = 2, X_2 = 1) + P(X_1 = 2, X_2 = 2)}{P(X_2 = 1) + P(X_2 = 2)} = \frac{0.05 + 0.05}{0.3 + 0.3} = \frac{1}{6} \quad \blacktriangle$$

Example 3.19   Roll a black die and a white die. Let X be the number on the black die and let Y be the sum of the numbers on the two dice. Let us determine the distribution of (X, Y) as well as the marginal distributions of X and Y.

We want to evaluate $P(X=i, Y=j)$ for $1 \le i \le 6$ and $2 \le j \le 12$. Now, $P(X=i, Y=j) = P(X=i, Y-X=j-i) = 1/36$ for $1 \le i < j \le 12$ and such that $j - i \le 6$, and 0 otherwise, because $Y - X$ represents the number on the white die, and thus a combination of i and j is either impossible for X and Y, yielding zero for the corresponding probability, or if it is possible, it then represents a single outcome out of the 36 possibilities for two dice.

Next, when evaluating $P(X=i)$, we note that $P(X=i, Y=j) = 0$ unless $i < j \le 6 + i$. That is, in

$$P(X=i) = \sum_{j=2}^{12} P(X=i, Y=j)$$

exactly six terms are different from zero, each of which is 1/36, yielding $P(X=i) = 1/6$. On the other hand,

$$P(Y=j) = \sum_{i=1}^{6} P(X=i, Y=j)$$

in which the nonzero terms, each of which is 1/36, are those for which either $1 \le i < j \le 6$ or $1 \le j - 6 \le i \le 6$. Hence

$$P(Y=j) = \begin{cases} \dfrac{j-1}{36} & \text{if } j \le 6 \\ \dfrac{13-j}{36} & \text{if } j > 6 \end{cases} \qquad \blacktriangle$$

We now go back to nondiscrete random vectors and look at some examples concerning distribution functions.

Example 3.20 The bivariate distribution function of (X_1, X_2) is

$$F(x, y) = 1 - e^{-x} - e^{-y} + e^{-(1/2)\max(x, y)}, \qquad x, y \ge 0$$

and $F(x, y) = 0$ otherwise. Let us determine the distribution functions of X_1 and X_2.

By letting $y \to +\infty$, we get

$$F_1(x) = P(X_1 \le x) = \lim F(x, y) = 1 - e^{-x} \qquad x \ge 0$$

Similarly, as $x \to +\infty$,

$$F_2(y) = P(X_2 \le y) = \lim F(x, y) = 1 - e^{-y} \qquad y \ge 0 \qquad \blacktriangle$$

Several multivariate distribution functions can lead to the same marginal distribution functions. This is seen from the following example when compared with Example 3.20.

Example 3.21   The bivariate distribution function of (X_1, X_2) is

$$F(x, y) = 1 - e^{-x} - e^{-y} + (e^x + e^y - 1)^{-1} \qquad x, y \geq 0$$

and it is 0 otherwise. Then the marginal distributions of $F(x, y)$ are again exponential.

Indeed, first letting $y \to +\infty$, we get $P(X_1 \leq x) = 1 - e^{-x}$, $x \geq 0$, and similarly, if $x \to +\infty$ in $F(x, y)$, the marginal $P(X_2 \leq y) = 1 - e^{-y}$, $y \geq 0$, is obtained.   ▲

If we compare $F(x, y)$ in Examples 3.20 and 3.21, we can recognize a partial similarity (both are given as the sum of four terms, the first three of which are identical in the two examples), and a substantial difference between them, namely, in terms of differentiability. The similarity of the two functions is due to (3.19) below.

Theorem 3.6   If (X_1, X_2) is a random vector with distribution function $F(x, y)$ and with marginals $F_1(x)$ and $F_2(y)$, then

$$(3.19) \qquad F(x, y) = F_1(x) + F_2(y) - 1 + P(X_1 > x, X_2 > y)$$

Furthermore, for any $a < b$ and $c < d$,

$$(3.20) \quad P(a < X_1 \leq b, c < X_2 \leq d) = F(b, d) - F(b, c) - F(a, d) + F(a, c)$$

Proof: To prove (3.19), define $A = \{X_1 > x\}$ and $B = \{X_2 > y\}$. Then $F(x, y) = P(A^c \cap B^c)$, $P(A) = 1 - F_1(x)$, and $P(B) = 1 - F_2(y)$. Therefore, by repeated application of Theorem 1.2,

$$F(x, y) = P(B^c) - P(A \cap B^c) = 1 - P(B) - [P(A) - P(A \cap B)]$$
$$= F_2(y) - [1 - F_1(x)] + P(X_1 > x, X_2 > y)$$

which is indeed (3.19).

Next, define $A_1 = \{X_1 \leq a\}$, $B_1 = \{X_1 \leq b\}$, $C = \{X_2 \leq c\}$, and $D = \{X_2 \leq d\}$. Then

$$P(a < X_1 \leq b, c < X_2 \leq d) = P(A_1^c \cap B_1 \cap C^c \cap D)$$

and thus, by Theorem 1.2 again,

$$P(a < X_1 \leq b, c < X_2 \leq d) = P(B_1 \cap C^c \cap D) - P(A_1 \cap B_1 \cap C^c \cap D) =$$
$$[P(B_1 \cap D) - P(B_1 \cap C \cap D)] - [P(A_1 \cap B_1 \cap D) - P(A_1 \cap B_1 \cap C \cap D)]$$

However, $A_1 \subset B_1$ and $C \subset D$ [i.e., $A_1 \cap B_1 = A_1$ and $C \cap D = C$], which, in view of the definition of A_1, B_1, C, and D, reduces the last formula to (3.20). This completes the proof.   ▲

Note that in (3.20) the sign of F is positive if both variables are either upper bounds (b, d) or both are lower bounds (a, c); in the mixed cases (a, d) and (b, c), the sign of F is negative.

The last term in (3.19) is called the *survival function* of (X_1, X_2), a concept that has a self-evident extension to higher dimension as well. The relation between distribution functions and survival functions in dimensions higher than 2, however, is quite complicated.

Example 3.22 The distribution function of (X_1, X_2),

$$F(x, y) = \begin{cases} 1 - e^{-2x} & \text{if } x > 0, \ y \geq 1 \\ (1 - e^{-2x})y^2 & \text{if } x > 0, \ 0 \leq y < 1 \\ 0 & \text{otherwise} \end{cases}$$

Let us evaluate $P(1/2 < X_1 \leq 3, \ 1/4 < X_2 \leq \leq 2)$.

By (3.20), the desired probability equals

$$F(3, 2) - F(1/2, 2) - F(3, 1/4) + F(1/2, 1/4)$$

Now, $F(3, 2)$ and $F(1/2, 2)$ are computed from $F(x, y) = 1 - e^{-2x}$, while $F(3, 1/4)$ and $F(1/2, 1/4)$ from $F(x, y) = (1 - e^{-2x})y^2$. Hence the answer is

$$(1 - e^{-6}) - (1 - e^{-1}) - (1 - e^{-6})\left(\frac{1}{4}\right)^2 + (1 - e^{-1})\left(\frac{1}{4}\right)^2 = \frac{15}{16}(e^{-1} - e^{-6}) \ \blacktriangle$$

Example 3.23 The bivariate function

$$F(x, y) = \begin{cases} 1 - e^{-2x} & \text{if } x > 0, \ y \geq 1 \\ \frac{1}{2}(1 - e^{-x}) & \text{if } x > 0, \ 0 \leq y < 1 \\ 0 & \text{otherwise} \end{cases}$$

has the properties (i) to (iii) of bivariate distribution functions as well as its marginals are univariate distribution functions. Yet $F(x, y)$ is not a distribution function.

Indeed, one can easily check the validity of the claims about properties (i) to (iii) of bivariate distribution functions and the marginals. However, should $F(x, y)$ be a distribution function, then, by (3.20), for every $a < b$ and $c < d$,

$$F(b, d) - F(b, c) - F(a, d) + F(a, c) \geq 0$$

But with $a = \log 4$, $b = \log 6$, $c = 1/2$, and $d = 2$, say (there are infinitely many choices), the expression above becomes

$$\left(1-\frac{1}{36}\right) - \frac{1}{2}\left(1-\frac{1}{6}\right) - \left(1-\frac{1}{16}\right) + \frac{1}{2}\left(1-\frac{1}{4}\right) = -\frac{1}{144} \quad \blacktriangle$$

This example shows that much care is to be taken if one wants to "construct" (rather than to compute) multivariate distribution functions.

We now return to our second remark on Examples 3.20 and 3.21, namely, differentiability of $F(x,y)$.

Definition 3.6   Let $(X_1,X_2,...,X_d)$ be a random vector with distribution function $F(z_1,z_2,...,z_d)$. If the d–fold partial derivative

$$(3.21) \qquad f(z_1,z_2,...,z_d) = \frac{\partial^{(d)}F(z_1,z_2,...,z_d)}{\partial z_1 \, \partial z_2 \cdots \partial z_d}$$

exists (except perhaps at a finite number of points), and if the order of differentiation is immaterial, $F(z_1,z_2,...,z_d)$ is called *absolutely continuous*, and $f(z_1,z_2,...,z_d)$ is called the *density function* of the vector $(X_1,X_2,...,X_d)$, or of $F(z_1,z_2,...,z_d)$.

It is customary to prove in calculus that the d-fold integral of the derivative f gives the function F:

$$(3.22) \quad F(z_1,z_2,...,z_d) = \int_{-\infty}^{z_1} \int_{-\infty}^{z_2} \cdots \int_{-\infty}^{z_d} f(u_1,u_2,...,u_d) \, du_1 \, du_2 \cdots du_d$$

In fact, more is true. Since the differenciation (3.21) actually expresses the probability that the vector $(X_1,X_2,...,X_d)$ is "very close" to the point $(z_1,z_2,...,z_d)$, integration of the density over a set A expresses (adds up) the probability that the vector $(X_1,X_2,...,X_d)$ falls into A:

$$(3.23) \quad P\{(X_1,X_2,...,X_d) \in A\} = \int \int \cdots \int_A f(u_1,u_2,...,u_d) \, du_1 \, du_2 \cdots du_d$$

When A is the intersection of the sets $\{-\infty < u_j \le z_j\}$, $1 \le j \le d$, (3.23) becomes (3.22).

Example 3.24   The joint density function of (X_1,X_2),

$$f(x,y) = \begin{cases} c & \text{if } 0 \le x \le y \le 1 \\ 0 & \text{otherwise} \end{cases}$$

where c is a constant. Let us determine the (marginal) distribution and density of both X_1 and X_2. Furthermore, let us evaluate the probabilities $P(X_1 \le 3/4, X_2 \le 1/2)$ and $P(X_1 \le (1/2)X_2)$.

We apply (3.22) with $d = 2$. By letting $z_2 \to +\infty$, we get, on one hand, the distribution function of X_1, while on the other, the double integral

$$(3.24) \quad F_1(z_1) = P(X_1 \leq z_1) = \int_{-\infty}^{z_1} \int_{-\infty}^{+\infty} f(u,v) \, du \, dv = \int_0^{z_1} \int_0^1 f(u,v) \, du \, dv$$

$$= \int_0^{z_1} \left(\int_u^1 c \, dv \right) du = c \int_0^{z_1} (1-u) \, du = c \left(z_1 - \frac{1}{2} z_1^2 \right)$$

where $0 \leq z_1 < 1$. Evidently, $F_1(z_1) = 0$ if $z_1 < 0$ and $F_1(z_1) = 1$ if $z_1 \geq 1$. Similarly to (3.24),

$$(3.25) \quad F_2(z_2) = P(X_2 \leq z_2) = \int_{-\infty}^{+\infty} \int_{-\infty}^{z_2} f(u,v) \, du \, dv = \int_0^{z_2} \left(\int_0^v c \, du \right) dv$$

$$= \frac{1}{2} c z_2^2, \qquad 0 \leq z_2 < 1$$

and $F_2(z_2) = 0$ or 1 according as $z_2 < 0$ or $z_2 \geq 1$.

Because the distribution functions $F_1(z_1)$ and $F_2(z_2)$ are continuous by definition, $1 = F_1(1) = c(1 - 1/2)$, or $1 = F_2(1) = 1/2c$, yields $c = 2$. The marginal density functions are obtained by differentiation; we have

$$f_1(z_1) = 2 - 2z_1 \text{ for } 0 < z_1 < 1, \text{ and } 0 \text{ otherwise}$$
$$f_2(z_2) = 2z_2 \text{ for } 0 < z_2 < 1, \text{ and } 0 \text{ otherwise}$$

From (3.22), and in view of $c = 2$,

$$P\left(X_1 \leq \frac{3}{4}, X_2 \leq \frac{1}{2} \right) = F\left(\frac{3}{4}, \frac{1}{2} \right) = \int_{-\infty}^{3/4} \int_{-\infty}^{1/2} f(z_1,z_2) \, dz_1 \, dz_2$$

$$= \int_0^{1/2} \left(\int_0^{z_2} 2 \, dz_1 \right) dz_2 = \int_0^{1/2} 2z_2 \, dz_2 = \frac{1}{4}$$

where the value 3/4 played a role to the extent only that it exceeds 1/2, that is, the structure of the density $f(z_1,z_2)$, reflecting the probability of (X_1,X_2) falling into a small neighborhood of (z_1,z_2), implies that $X_1 < X_2$ (with probability 1).

Finally, by utilizing (3.23), we have

$$P\left(X_1 \leq \frac{1}{2} X_2 \right) = \int_0^1 \int_0^{(1/2)z_2} 2 \, dz_1 \, dz_2 = \int_0^1 z_2 \, dz_2 = \frac{1}{2} \qquad \blacktriangle$$

The first integral formulas of (3.24) and (3.25) are valid whatever the density function. Hence, by differentiation we get that if (X_1,X_2) is a random vector with density function $f(u,v)$, then

$$(3.26) \qquad f_1(z_1) = \int_{-\infty}^{+\infty} f(z_1,v) \, dv \qquad \text{(density function of } X_1)$$

and

(3.27) $f_2(z_2) = \int_{-\infty}^{+\infty} f(u,z_2)\, du$ (density function of X_2)

Formulas similar to (3.24) to (3.27) are valid in dimensions higher than two as well.

There are very few bivariate density functions of wide acceptance in the literature. One of them is the (bivariate) normal density function, defined as

(3.28) $f(x,y) = \dfrac{1}{2\pi\sigma_1\sigma_2\sqrt{1-\rho^2}}\, e^{-(1/2)Q(x,y)}$ $-\infty < x, y < +\infty$

where $\sigma_1 > 0$, $\sigma_2 > 0$, $-1 < \rho < 1$, and

(3.29) $Q(x,y) = \dfrac{1}{1-\rho^2}\left[\dfrac{(x-m_1)^2}{\sigma_1^2} - 2\rho\dfrac{x-m_1}{\sigma_1}\dfrac{y-m_2}{\sigma_2} + \dfrac{(y-m_2)^2}{\sigma_2^2}\right]$

with some real numbers m_1 and m_2.

Several properties of this density are established in subsequent sections. The reader is advised to apply (3.26) and (3.27) to conclude that the marginal densities of the bivariate normal density function (3.28) – (3.29) are the normal densities

(3.30) $f_i(z_i) = \dfrac{1}{\sqrt{2\pi}\sigma_i}\, e^{-(z_i - m_i)^2/2\sigma_i^2}$ $i = 1, 2$

3.8   CONDITIONAL DENSITIES

If Y is a discrete random variable taking on the values $y_1, y_2, \ldots$ with $P(Y = y_j) = p_j > 0$, then, for a random variable X, the conditional distribution

(3.31) $F(z \mid y) = P(X \le z \mid Y = y)$

where $y = y_j$, $j \ge 1$, can be computed directly from the definition of conditional probabilities. On the other hand, if the distribution function of Y is continuous, then $P(Y = y) = 0$ for all y (Theorem 3.3), and thus (3.31) has not been defined so far. However, practical problems such as the need to know the distribution function of a random service time, given that a (random) part of the service has been completed, or conclusions in laboratories, given that some (random) measurements have been taken, require a method for computing the conditional distribution function (3.31). We give below a definition that is restricted to the case when (X, Y) has density.

Definition 3.7   Assume that the random vector (X, Y) has a density function $f(x, y)$. Then

(3.32) $F(z\,|\,y) = P(X \le z\,|\,Y = y) = \lim_{\triangle y = 0} P(X \le z\,|\,y < Y \le y + \triangle y)$

whenever the density function $f_Y(y)$ of Y is positive.

Theorem 3.7   Let $f(x,y)$ be the density function of the vector (X,Y). Assume that Y is such that the density function $f_Y(y)$ of Y is positive. Then the conditional distribution function $F(z\,|\,y)$ defined in (3.32) becomes

(3.33) $F(z\,|\,y) = \dfrac{1}{f_Y(y)} \displaystyle\int_{-\infty}^{z} f(x,y)\,dx$

or equivalently, the conditional density function $f(z\,|\,y)$, defined as the derivative of $F(z\,|\,y)$ with respect to z, satisfies

(3.34) $f(z,y) = f(z\,|\,y)f_Y(y)$

   Proof: By the definition of conditional probabilities,

(3.35) $P(X \le z\,|\,y < Y \le y + \triangle y) = \dfrac{P(X \le z, y < Y \le y + \Delta y)}{P(y < Y \le y + \Delta y)}$

Define $A = \{Y \le y\}$ and $B = \{X \le z, Y \le y + \triangle y\}$. Then the numerator of the right-hand side of (3.35) is $P(A^c \cap B)$, which, by Theorem 1.2, can also be written as

   $P(B) - P(A \cap B) = P(X \le z, Y \le y + \triangle y) - P(X \le z, Y \le y)$

Hence, denoting by $F(x,y)$ and $F_Y(y)$, respectively, the distribution function of (X,Y) and of Y, the right-hand side of (3.35) equals

$$\frac{F(z,y + \triangle y) - F(z,y)}{F_Y(y + \triangle y) - F_Y(y)} = \frac{[F(z,y + \triangle y) - F(z,y)]/\triangle y}{[F_Y(y + \triangle y) - F_Y(y)]/\triangle y}$$

Upon letting $\triangle y \to 0$, the limit of this last ratio, combined with (3.32), yields

$$F(z\,|\,y) = \frac{1}{f_Y(y)} \frac{\partial F(z,y)}{\partial y}$$

from which, by differentiation with respect to z, we get (3.34). Formula (3.34), on the other hand, by integration with respect of z, leads to (3.33). This completes the proof.   ▲

   Note that the combination of (3.26) and (3.34) gives

(3.36) $f_X(z) = \displaystyle\int_{-\infty}^{+\infty} f(z\,|\,y)f_Y(y)\,dy$ (density of X)

where $f(z\,|\,y)f_Y(y) = 0$ if $f_Y(y) = 0$ even though $f(z\,|\,y)$ is not defined in such a case. Formula (3.36) can be viewed as the continuous version of the total probability rule (Theorem 2.3).

Example 3.25   Let the random variable Y be uniformly distributed on $(0,1)$ [i.e., $f_Y(y) = 1$ for $0 < y < 1$, and 0 otherwise]. Assume that given $Y = y$, $0 < y < 1$, X is exponentially distributed with density function

$$f(z \mid y) = ye^{-yz} \qquad z \geq 0$$

(and 0 for $z < 0$). Let us determine the distribution function of X.
    Utilizing (3.36), and integrating by parts, we have

$$f_X(z) = \int_0^1 ye^{-yz}\,dy = -y\frac{e^{-yz}}{z}\bigg]_{y=0}^1 + \int_0^1 \frac{e^{-yz}}{z}\,dy$$

$$= \frac{1 - e^{-z} - ze^{-z}}{z^2} \qquad z \geq 0$$

and $f_X(z) = 0$ for $z < 0$. Hence $F_X(z) = 0$ if $z < 0$, and

$$F_X(z) = \int_0^z \frac{1 - e^{-u} - ue^{-u}}{u^2}\,du = 1 - \frac{1 - e^{-z}}{z} \qquad z \geq 0 \qquad \blacktriangle$$

Example 3.26   Select two points X and Y as follows. First X is selected uniformly on $(0,1)$. Then, given $X = x$, Y is selected uniformly on the interval $(0,x)$. What is the probability of $X \geq 1/2$, given $Y = 1/4$?
    By the definition of uniform distribution, the density of X,

$$f_X(x) = \begin{cases} 1 & \text{if } 0 < x < 1 \\ 0 & \text{otherwise} \end{cases}$$

and the conditional density of Y, given $X = x$,

$$f(y \mid x) = \begin{cases} \dfrac{1}{x} & \text{if } 0 < y < x \\ 0 & \text{otherwise} \end{cases}$$

Therefore, the density of the vector (X,Y), computed by (3.34),

$$f(x,y) = \begin{cases} \dfrac{1}{x} & \text{if } 0 < y < x < 1 \\ 0 & \text{otherwise} \end{cases}$$

Let us now compute the density of Y by an appeal to (3.36). We get

$$f_Y(y) = \int_0^1 f(y \mid x)f_X(x)\,dx = \int_y^1 \frac{1}{x}\,dx = -\log y \qquad 0 < y < 1$$

Hence, by (3.33),

$$P\left(X \ge \frac{1}{2} \,\middle|\, Y = \frac{1}{4}\right) = 1 - P\left(X < \frac{1}{2} \,\middle|\, Y = \frac{1}{4}\right) = 1 - \frac{1}{f_Y(1/4)} \int_0^{1/2} f\left(x, \frac{1}{4}\right) dx$$

$$= 1 - \frac{1}{\log 4} \int_{1/4}^{1/2} \frac{1}{x}\, dx = 1 - \frac{\log 4 - \log 2}{\log 4} = \frac{1}{2} \qquad \blacktriangle$$

3.9 INDEPENDENCE OF RANDOM VARIABLES

Definition 3.8 The random variables X_1, X_2, ..., X_n are said to be independent if their n-dimensional distribution function $F(z_1, z_2, \ldots, z_n)$ splits into the product of the univariate marginal distribution functions $F_j(z_j) = P(X_j \le z_j)$, that is,

$$(3.37) \qquad F(z_1, z_2, \ldots, z_n) = F_1(z_1) F_2(z_2) \cdots F_n(z_n)$$

for all real numbers z_j, $1 \le j \le n$.

Note that by letting one, or several, $z_j \to +\infty$, the corresponding $F_j(z_j) \to 1$, and the left-hand side of (3.37) converges to the corresponding lower-dimensional marginal distribution. Hence, if X_1, X_2, ..., X_n are independent, so are all subsets of them [in other words, if n random variables are independent, (3.37) applies to all lower-dimensional marginals]. This observation thus yields: The random variables X_1, X_2, ..., X_n are independent if, and only if, for all real numbers z_j, $1 \le j \le n$, the events $\{X_j \le z_j\}$, $1 \le j \le n$, are independent.

With $n = 2$, (3.20) and the definition (3.37) imply that, for independent random variables X_1 and X_2, and for all numbers $a < b$ and $c < d$,

$$P(a < X_1 \le b,\ c < X_2 \le d)$$
$$= F_1(b)F_2(d) - F_1(b)F_2(c) - F_1(a)F_2(d) + F_1(a)F_2(c)$$
$$= [F_1(b) - F_1(a)]\,[F_2(d) - F_2(c)]$$
$$= P(a < X_1 \le b)P(c < X_2 \le d)$$

that is, the events $\{a < X_1 \le b\}$ and $\{c < X_2 \le d\}$ are independent. With an argument similar to the one leading to (3.20), we can prove that this result extends to arbitrary n: For independent random variables X_1, X_2, ..., X_n, and for arbitrary numbers $a_j < b_j$, $1 \le j \le n$, the events $\{a_j < X_j \le b_j\}$, $1 \le j \le n$, are independent.

When the random variables X_j are discrete, it is more convenient to deal with their distribution than with distribution functions. From the definition of independence itself, or from the property stated in the preceding paragraph, it follows that the discrete random variables X_1, X_2, ..., X_n are independent if for all possible values x_{j,m_j} of X_j, $1 \le j \le n$, the events $\{X_j = x_{j,m_j}\}$, $1 \le j \le n$, are independent, that is,

$$P(X_j = x_{j,m_j}, 1 \le j \le n) = \prod_{j=1}^{n} P(X_j = x_{j,m_j})$$

Example 3.27   What is the probability that two persons have to roll a die the same number of times to get a 6?

Let us call the two people A and B, and let X and Y be the number of times, A and B, respectively, had to roll the die in order to get a 6. Since A and B do not influence each other's outcome, X and Y are independent. The distribution of both X and Y is geometric with parameter $p = 1/6$ (recall the discrete waiting-time model from Section 3.4). Hence the desired probability

$$P(X = Y) = \sum_{j=1}^{+\infty} P(X = j, Y = j) = \sum_{j=1}^{+\infty} P(X = j)P(Y = j)$$

$$= \sum_{j=1}^{+\infty} \left[\frac{1}{6} \left(\frac{5}{6} \right)^{j-1} \right]^2 = \left(\frac{1}{6} \right)^2 \frac{1}{1 - (5/6)^2} = \frac{1}{11}$$

where the last infinite sum was computed by an appeal to (1.25).   ▲

Assume now that the density of F in (3.37) exists. Then, by n-fold (partial) differentiation we get from (3.37) that, for independent random variables X_1, X_2, ..., X_n with joint density function $f(z_1, z_2, ..., z_n)$ and with marginal density functions $f_j(z_j)$,

(3.38) $f(z_1, z_2, ..., z_n) = f_1(z_1)f_2(z_2) \cdots f_n(z_n)$

Conversely, by integrating (3.38), we get (3.37); that is, (3.37) and (3.38) are equivalent assumptions. Note that, by (3.34) and (3.38), for independent random variables X and Y, $f(z \mid y) = f_1(z)$, where $f(z \mid y)$ is the conditional density of X, given $Y = y$, and $f_1(z)$ is the unconditional density function of X.

Example 3.28   The density function of the vector (X, Y),

$$f(x,y) = \begin{cases} xe^{-y} & \text{if } 0 < x < 1 \text{ and } y > 0 \\ 0 & \text{otherwise} \end{cases}$$

Consequently, X and Y are independent.

Indeed, one immediately finds that (3.38) holds [and that X is uniform on $(0, 1)$ and Y is exponential].   ▲

Example 3.29 The density of the vector (X,Y),

$$f(x,y) = \begin{cases} 8xy & \text{if } 0 < x < y < 1 \\ 0 & \text{otherwise} \end{cases}$$

Hence the random variables X and Y are dependent.

At first glance it looks as if $f(x,y)$ would again be the product of two functions, one with variable x and the other with y, suggesting that (3.38) might hold. However, the domain $0 < x < y < 1$ shows that it is not the case. Indeed, by (3.26), for $0 < x < 1$,

$$f_X(x) = \int_0^1 f(x,y)\, dy = 8 \int_x^1 xy\, dy = 4x(1 - x^2)$$

and, by (3.27), for $0 < y < 1$,

$$f_Y(y) = \int_0^1 f(x,y)\, dx = 8 \int_0^y xy\, dx = 4y^3$$

and thus (3.38) fails. ▲

Example 3.30 Let the random variables X_1, X_2, and X_3 be independent and uniformly distributed on $(0, 1)$. Let W be the smallest and Z be the largest of X_1, X_2, and X_3. Let us evaluate (i) $P(W > 1/3)$; (ii) $P(Z \le 1/2)$; and (iii) $P(1/3 < W$ and $Z \le 1/2)$.

By assumption, each X_j has the distribution function $F(z) = z$ for $0 < z < 1$. Now, since

$$\left\{ W > \frac{1}{3} \right\} = \left\{ X_1 > \frac{1}{3},\ X_2 > \frac{1}{3},\ X_3 > \frac{1}{3} \right\},$$

$$\left\{ Z \le \frac{1}{2} \right\} = \left\{ X_1 \le \frac{1}{2},\ X_2 \le \frac{1}{2},\ X_3 \le \frac{1}{2} \right\}$$

and

$$\left\{ \frac{1}{3} < W \text{ and } Z \le \frac{1}{2} \right\} = \left\{ \frac{1}{3} < X_1 \le \frac{1}{2},\ \frac{1}{3} < X_2 \le \frac{1}{2},\ \frac{1}{3} < X_3 \le \frac{1}{2} \right\}$$

by independence,

$$P\left(W > \frac{1}{3} \right) = \left[1 - F\left(\frac{1}{3} \right) \right]^3 = \left(\frac{2}{3} \right)^3 = \frac{8}{27}, \qquad P\left(Z \le \frac{1}{2} \right) = F^3\left(\frac{1}{2} \right) = \frac{1}{8}$$

and

$$P\left(\frac{1}{3} < W \text{ and } Z \le \frac{1}{2} \right) = \left[F\left(\frac{1}{2} \right) - F\left(\frac{1}{3} \right) \right]^3 = \left(\frac{1}{2} - \frac{1}{3} \right)^3 = \frac{1}{216} \qquad ▲$$

Example 3.31   Let the components of the random point (X,Y) in the plane be independent, each with standard normal density function. Let us determine the distribution of the slope of the line connecting (X,Y) and the origin.

The slope of the line in question is Y/X. Its distribution function can be computed by (3.23), where $A = \{(x,y): y/x \leq z\}$ and the density of (X,Y),

$$f(x,y) = \frac{1}{2\pi} e^{-(1/2)(x^2 + y^2)} \qquad -\infty < x, y < +\infty$$

We thus have [we can ignore the possibility of $X = 0$, because $P(X = 0) = 0$ in view of the distribution being continuous],

$$P\left(\frac{Y}{X} \leq z\right) = \frac{1}{2\pi} \int\int_A e^{-(1/2)(x^2 + y^2)} \, dx \, dy$$

$$= \frac{1}{2\pi} \int_0^{+\infty} \left[\int_{-\infty}^{xz} e^{-(1/2)(x^2 + y^2)} \, dy\right] dx + \frac{1}{2\pi} \int_{-\infty}^0 \left[\int_{xz}^{+\infty} e^{-(1/2)(x^2 + y^2)} \, dy\right] dx$$

$$= \frac{1}{\sqrt{2\pi}} \int_0^{+\infty} e^{-(1/2)x^2} N(xz) \, dx + \frac{1}{\sqrt{2\pi}} \int_{-\infty}^0 e^{-(1/2)x^2}[1 - N(xz)] dx$$

where $N(u)$ is the standard normal distribution function. In view of $1 - N(xz) = N(-xz)$, with the substitution $x = -v$, the second integral can be brought into the form of the first one. In addition, when one writes $N(xz)$ in detail as an integral, with one additional simple substitution and by interchanging the order of integration, we get

$$P\left(\frac{Y}{X} \leq z\right) = \frac{1}{\pi} \int_{-\infty}^z \left[\int_0^{+\infty} xe^{-(1/2)(x^2 + v^2 x^2)} \, dx\right] dv$$

The inner integral is $1/(1 + v^2)$. Consequently,

$$P\left(\frac{Y}{X} \leq z\right) = \frac{1}{\pi} \int_{-\infty}^z \frac{1}{1 + v^2} dv = \frac{1}{2} + \frac{1}{\pi} \arctan z \qquad \blacktriangle$$

The distribution function obtained here is known as the *Cauchy distribution function*.

3.10   SUMS OF INDEPENDENT RANDOM VARIABLES

One of the most common problems faced in practice is to determine the distribution of sums of independent random variables. For example, if servicing the jth customer requires the random time X_j, then the sum $X_1 + X_2 + \cdots + X_n$ expresses the aggregate time of servicing n customers. Other examples frequently faced are the aggregate profit on n products; the number of patients registering at a hospital during a week in terms of those during the days of the week; and so on.

It evidently suffices to know a method for evaluating the distribution of the sum of two random variables. Such methods are given in the next two theorems.

Theorem 3.8   Let X and Y be independent discrete random variables, taking on the values $x_1, x_2, \ldots$ and $y_1, y_2, \ldots$, respectively, with distribution $P(X = x_i) = p_i$ and $P(Y = y_j) = r_j$. Then

$$P(X + Y = s) = \sum_{x_i + y_j = s} p_i r_j$$

Theorem 3.9   Let X and Y be independent random variables with density function $f(x)$ and $g(y)$, respectively. Then $Z = X + Y$ is absolutely continuous, whose density function

$$h(z) = \int_{-\infty}^{+\infty} f(x)g(z - x) \, dx$$

Because Z is symmetric in X and Y, one can evidently interchange the roles of f and g in the formula above.

Proof of Theorem 3.8:   The event $X + Y = s$ can be written as the union of as many mutually exclusive events of the form $\{X = x_i, Y = y_j\}$ as ways s can be written as $s = x_i + y_j$. If we now take probabilities, the probability of the union of exclusive terms becomes a sum, and, by independence, $P(X = x_i, Y = y_j) = p_i r_j$, which completes the proof.   ▲

Proof of Theorem 3.9:   We apply the general formula (3.23) in the form

$$P(X + Y \le z) = \iint_{x + y \le z} f(x)g(y) \, dx \, dy = \int_{-\infty}^{+\infty} \left[\int_{-\infty}^{z - x} f(x)g(y) \, dy \right] dx$$

We can see that the expression above is differentiable in z, and the actual differentiation gives the claimed formula for $h(z)$. The theorem is established.   ▲

Example 3.32   Assume that the number X of customers arriving at a station in any half-hour interval is either 0, or 1, or 2, with distribution $P(X = 0) = 0.1$, $P(X = 1) = 0.6$, and $P(X = 2) = 0.3$ (computed from past experience, say). What is the distribution of the number of customers arriving in a 1-hour interval?

One can assume that the number of customers arriving in nonoverlapping time intervals is independent. Hence the question above is about the distribution of $X_1 + X_2$, where X_1 and X_2 are independent, and each is distributed as X. We can thus apply Theorem 2.8 (with $p_i = r_i$). We have

$$P(X_1 + X_2 = 0) = 0.1^2 = 0.01$$
$$P(X_1 + X_2 = 1) = 0.1 \times 0.6 + 0.6 \times 0.1 = 0.12$$
$$P(X_1 + X_2 = 2) = 0.1 \times 0.3 + 0.6 \times 0.6 + 0.3 \times 0.1 = 0.42$$
$$P(X_1 + X_2 = 3) = 0.6 \times 0.3 + 0.3 \times 0.6 = 0.36$$
$$P(X_1 + X_2 = 4) = 0.3^2 = 0.09$$ ▲

Example 3.33 Let X and Y be independent Poisson variables with parameters λ and $s\mu$, respectively. Let us show that $Z = X + Y$ is also Poisson with parameter $\lambda + \mu$.

By assumption, for nonnegative integers i and j,

$$p_i = P(X = i) = \frac{\lambda^i e^{-\lambda}}{i!} \quad \text{and} \quad r_j = P(Y = j) = \frac{\mu^j e^{-\mu}}{j!}$$

Hence, by Theorem 3.8,

$$P(Z = s) = \sum_{i+j=s} p_i r_j = \sum_{i=0}^{s} p_i r_{s-i} = \sum_{i=0}^{s} \frac{\lambda^i e^{-\lambda}}{i!} \frac{\mu^{s-i} e^{-\mu}}{(s-i)!}$$
$$= \frac{e^{-\lambda} e^{-\mu} \mu^s}{s!} \sum_{i=0}^{s} \frac{(\lambda/\mu)^i s!}{i!(s-i)!}$$

Because $s!/[i!(s-i)!] = \binom{s}{i}$, the last sum above, by the binomial formula (2.22), equals $[1 + (\lambda/\mu)]^s$. Therefore,

$$P(Z = s) = \frac{e^{-(\lambda + \mu)}}{s!} (\lambda + \mu)^s$$

which is the claimed Poisson distribution.   ▲

Example 3.34 Assume that a doctor can be visited by appointment only, and patients are scheduled at 25-minute intervals. Assume that consultation consists of a 20-minute routine checkup and conversation, and a random time X, which is exponentially distributed with density

$$f(x) = \frac{1}{5} e^{-x/5} \qquad x > 0$$

and $f(x) = 0$ for $x < 0$, where X is measured in minutes. Let us evaluate (i) the probability that the doctor has 5 minutes free after the second patient, and (ii) the probability that the third patient has to wait at least 7 minutes.

The consultation with the second patient ends in $40 + X_1 + X_2$ minutes, where X_1 and X_2 are independent random variables, each being distributed as X. Both (i) and (ii) can be answered in terms of the distribution function of $X_1 + X_2$, which in turn can be computed from its density. By Theorem 3.9, the density of $Z = X_1 + X_2$,

$$h(z) = \int_{-\infty}^{+\infty} f(x)f(z-x)\, dx = \frac{1}{25} \int_0^z e^{-x/5} e^{-(z-x)/5}\, dx$$

$$= \frac{1}{25} z e^{-z/5} \qquad z > 0$$

and evidently, $h(z) = 0$ for $z < 0$ [note that the limits of the integral above became 0 and z, because, for $x < 0$, $f(x) = 0$, while for $x > z$, $f(z-x) = 0$]. Now, integrating by parts yields ($z > 0$)

$$H(z) = P(Z \le z) = \frac{1}{25} \int_0^z u e^{-u/5}\, du = 1 - e^{-z/5}\left(1 + \frac{1}{5}z\right)$$

Clearly, the desired probability in (i) equals

$$P(Z \le 5) = H(5) = 1 - 2e^{-1} = 0.26$$

and the answer to (ii) is

$$P(Z \ge 17) = 1 - H(17) = 4.4e^{-4.4} = 0.15 \qquad\qquad \blacktriangle$$

Example 3.35 Let X and Y be independent random variables, each with standard normal density function. Then $Z = X + Y$ is normally distributed.
    By Theorem 3.9, the density function of Z,

$$h(z) = \frac{1}{2\pi} \int_{-\infty}^{+\infty} e^{-(1/2)x^2} e^{-(1/2)(z-x)^2}\, dx$$

The combined exponents

$$-\frac{1}{2}x^2 - \frac{1}{2}(z-x)^2 = -\left(x - \frac{1}{2}z\right)^2 - \frac{1}{4}z^2$$

and thus

$$h(z) = \frac{1}{2\pi} e^{-(1/4)z^2} \int_{-\infty}^{+\infty} e^{-[x-(1/2)z]^2}\, dx$$

which, with the notation (3.14) of the general normal distribution function, becomes

$$h(z) = (4\pi)^{-1/2} e^{-(1/4)z^2} N\left(+\infty; \frac{1}{2}z, 2^{-1/2}\right) = (4\pi)^{-1/2} e^{-z^2/4}$$

because every distribution function at $+\infty$ equals 1. $\blacktriangle$

By utilizing (3.13), one can now easily deduce from Example 3.35 that sums of independent normal random variables (without the restriction of being standard normal) are normal.

3.11 METHODS FOR DETERMINING DISTRIBUTION FUNCTIONS

In previous sections the focus was on the development of methods for utilizing distribution functions in computations of probabilities involving random variables. Here we summarize the methods applied in practice for determining distribution functions. These fall into the following headings:
(i) From basic assumptions, the distribution function is uniquely determined by a mathematical argument (characterizations of distributions).
(ii) Approximations (either through mathematical limit theorems, or by empirical methods).
(iii) Statistical testing (called goodness-of-fit tests).

Evidently, the most desirable method for an applied scientist is the availability of a characterization theorem. In this way, there is no "second guessing," and the conclusion in such a model is very sound. Several results in the present book fall into this category. The most evident is the model of Section 3.3, where the exponential distribution is shown to be the only distribution function of random life terminated by an accident, or, more generally, when aging is not a factor in the length of "life." But others are equally important; for example, we proved that the frequency of an event in independent repetitions of an experiment is binomial; the distribution of type I items obtained in a selection without replacement is hypergeometric, and the discrete waiting time is geometric. We have also mentioned (whose proof is delayed until Chapter 5) that measurement errors due to inaccuracies of instruments are normally distributed. In Chapter 5 we shall also give some indications of why the strength of sheets of metals (measured by the length of time a sheet can be put under periodic stress without breaking) follows the *Weibull distribution*, defined as

$$F(z) = 1 - e^{-z^a}, \qquad a > 0, \quad z \geq 0$$

Note that the special case $a = 1$ is the exponential distribution. Later in this section we discuss the Poisson process, which has a wide variety of applications, and which shows that under its conditions, the continuous waiting time is exponentially distributed. In that model we also establish that the distribution of the number of occurrences of such events as the arrival of customers at a store, the breakdown of a piece of equipment, and so on, in a fixed time interval is Poisson. Although there are several characterization theorems not covered in the present book, it should be added that this field is relatively new. Much of the material in the book by J. Galambos and S. Kotz, *Characterizations of Probability Distributions*, (Springer Verlag, Berlin, 1978), can be understood by someone who has gone through the present book.

For most practical problems approximation methods of finding the distribution function of a random variable is sufficient, and in some cases, it is even superior to a characterization. For example, the normal approximation to the binomial distribution is the only way of computing probabilities concerning the frequency of an event when the parameter n of the binomial distribution is large. Another approximation theorem is the Poisson approximation to the binomial (Section 2.10). Both of these approximations, however, are for facilitating computations in an exact model. More significant from the point of view of the present discussion is when a limit theorem is applied to replace a distribution function that we do not know. The most remarkable of all theorems of this kind is the central limit theorem, discussed in detail in Chapter 5. It says that averages of a large number of independent random variables with the same distribution function are approximately normal. Therefore, whenever we face a large number of copies of the same phenomenon (such as the amounts of claim at an insurance company, profit of items produced, amount in the bank accounts of customers, etc.), we do not always gain by the knowledge of the distribution of the individual terms, because the approximate normality of the aggregates will be utilized after all (the major concern of a bank is the fluctuation of the sum of all deposits rather than the fluctuation of individual deposits). Several other limit theorems are presented in Chapters 5 and 6.

When neither a characterization theorem nor a general mathematical limit theorem is available to the applied scientist, the distribution function has to be approximated through empirical studies, which can be through a large number of actual experiments or by simulation on a computer. The method is as follows. Assume that we would like to get an approximation to the distribution function $F(z)$ of the random variable X. We observe X n times, independently of each other, yielding $X_1, X_2, ..., X_n$. For every z, we count the number $k_z(n)$ of those X_j which satisfy $X_j \le z$. Then

$$F_n(z) = \frac{1}{n} k_z(n)$$

called the *empirical distribution function* (of X, based on the observations $X_1, X_2, ..., X_n$), converges to $F(z)$ by the Chebyshev inequality (Theorem 2.9). As a matter of fact, putting $A = \{X \le z\}$, then $P(A) = F(z)$, and $F_n(z)$ is exactly the relative frequency of A in the n repetitions concerning X (observations on X). [Because of this important relation of $F_n(z)$ to $F(z)$, the approximation of $F(z)$ by $F_n(z)$ has been investigated in great detail in the literature.] If it is not too expensive to get a large number n of observations on X, this empirical approximation can be made quite accurate. (However, the expense of doing so can be a major factor, such as in space research, and an additional limitation relates to safety considerations, such as in medical and nuclear research.)

Although in some cases approximation methods are adequate or even superior to characterizations, there is a significant group of practical problems for which they must be avoided. Common to this group is the fact that they are related to safety and, at the same time, to extremal properties. As an example, let us look at the problem of floods on a river. At a given point Z of the river C, the depth X of the water is random with distribution function $F(z)$. Depending on Z and C, there is a value u such that if $X > u$, we have flooding. Hence, if X_j is the water level (of C at Z) on the jth day of the year, then in order to compute the probability of having no flooding in a year, we need the value

$$P(X_1 \le u, X_2 \le u, \ldots, X_{365} \le u) = F^{365}(u)$$

where we assumed that the X_j are independent. Now, if the true value $F(u) = 0.999$, then $F^{365}(u) = 0.69$. However, if $F(u)$ is replaced by an approximation $G(u)$, say, which gives $G(u) = 0.9999$, then $G^{365}(u) = 0.96$ is the value used in any decision. This value suggests that there is practically no danger of flooding for the next 10 years, whereas the true value (0.69) practically assures a flood in 10 years (in fact, three floods can be anticipated).

As a last resort, one can use other statistical methods, called goodness-of-fit tests, to "determine" the distribution function of a random variable. This method consists of two steps: first, a distribution function is assumed to be the true one (unfortunately, often for no reason whatsoever as to the particular choice), and then observations are used to test whether we should look for a reason to change the assumption. That is, observations are not used to confirm the correctness of the assumption; rather, the data are challenged to contradict it.

It should be clear by now why contradictory statements are made by different scientists, even though they use the same data, combined with the same theorems from probability theory and statistics. They differ at the starting point.

We conclude this section with an important model.

The Poisson Process

We want to determine the distribution of the (random) number of times a specific event occurs in a given time interval (a, b). For example, an event may be that a customer arrives at a store, that a specific instrument breaks down, that a deposit in excess of $5000 is made at a bank, and so on. In the formulation of the model below, we use the term "number of points" in place of "number of occurrences."

Definition 3.9   We say that points are entered on the positive real line according to a Poisson process if the (random) number $X(a, b)$ of points in the interval (a, b) satisfies the following conditions:
(i) $X(a, b)$ and $X(c, d)$ are independent whenever (a, b) and (c, d) do not overlap.
(ii) The distribution of $X(a, a + t)$, $t > 0$, is the same as that of $X(t) = X(0, t)$ for all $a > 0$.
(iii) There is a number $\lambda > 0$ such that

$$\lim_{t = 0} \frac{P(X(t) = 1) - \lambda t}{t} = 0$$

(iv) $\displaystyle\lim_{t = 0} \frac{P(X(t) \geq 2)}{t} = 0$

Note the meaning of the assumptions. Conditions (i) and (ii) express the fact that the starting point of the process is immaterial; in additon, every segment of the process is independent of the past as well as of the future. Conditions (iii) and (iv), on the other hand, describe the structure of the process in small intervals. They say that the probability that exactly one point will be in a small interval is proportional to the length of the interval, while the probability of having more than one point in very small intervals is "unlikely."

Theorem 3.10   In a Poisson process, for every fixed $t > 0$,

$$P(X(t) = k) = \frac{(\lambda t)^k e^{-\lambda t}}{k!} \qquad k = 0, 1, 2, \ldots$$

Corollary   Let Y be the time until the first point enters in a Poisson process. Then

$$P(Y \leq t) = 1 - e^{-\lambda t} \qquad t > 0$$

Proof of the Corollary: Because $\{Y > t\} = \{X(t) = 0\}$, the distribution of $X(t)$ with $k = 0$ gives the exponential distribution claimed for Y.   ▲

Proof of Theorem 3.10:   Set $p_k(t) = P(X(t) = k)$. The proof is based on decomposing the sets $\{X(t + \triangle t) = k\}$ into unions of sets of the kind

$$A_{i,j} = \{X(t) = i\} \cap \{X(t, t + \triangle t) = j\}$$

Note that, by assumptions (i) and (ii),

(3.39) $$P(A_{i,j}) = p_i(t)p_j(\triangle t)$$

Let us also rewrite the assumptions in terms of $p_k(t)$. First, (iii) becomes

(3.40) $$\frac{p_1(\triangle t)}{\triangle t} \to \lambda \qquad \text{as } \triangle t \to 0$$

and since $X(t)$ either equals 0 or 1, or $X(t) \geq 2$, the combination of (iii) and (iv) yields

(3.41) $$\frac{1 - p_0(\triangle t)}{\triangle t} \to \lambda \qquad \text{as } \triangle t \to 0$$

Now, the mentioned decompositions are as follows:

$$\{X(t + \triangle t) = 0\} = A_{0,0}; \qquad \{X(t + \triangle t) = 1\} = A_{1,0} \cup A_{0,1}$$

and for $k \geq 2$,

$$\{X(t + \triangle t) = k\} = A_{k,0} \cup A_{k-1,1} \cup B_k$$

where

$$B_k = \{X(t) \leq k - 2\} \cap \{X(t, t + \triangle t) \geq 2\}$$

and thus

(3.42) $$P(B_k) \leq P(X(t, t + \triangle t) \geq 2) = P(X(\triangle t) \geq 2)$$

Taking the probabilities above, in view of (3.39), yields

$$p_0(t + \triangle t) = p_0(t)p_0(\triangle t)$$
$$p_1(t + \triangle t) = p_1(t)p_0(\triangle t) + p_0(t)p_1(\triangle t)$$

and for $k \geq 2$,

$$p_k(t + \triangle t) = p_k(t)p_0(\triangle t) + p_{k-1}(t)p_1(\triangle t) + P(B_k)$$

Next, we subtract from these equations $p_0(t)$, $p_1(t)$, and $p_k(t)$, respectively, and divide each by $\triangle t$; we get

(3.43) $$\frac{p_0(t + \triangle t) - p_0(t)}{\triangle t} = p_0(t)\frac{p_0(\triangle t) - 1}{\triangle t}$$

(3.44) $$\frac{p_1(t + \triangle t) - p_1(t)}{\triangle t} = p_1(t)\frac{p_0(\triangle t) - 1}{\triangle t} + p_0(t)\frac{p_1(\triangle t)}{\triangle t}$$

and for $k \geq 2$,

(3.45) $$\frac{p_k(t + \triangle t) - p_k(t)}{\triangle t} = p_k(t)\frac{p_0(\triangle t) - 1}{\triangle t} + p_{k-1}(t)\frac{p_1(\triangle t)}{\triangle t} + \frac{P(B_k)}{\triangle t}$$

We shall solve the equations (3.43) to (3.45) sequentially by letting $\triangle t \to 0$.

By (3.41), the right-hand side of (3.43) has a limit; therefore, so does the left-hand side, giving

$$\frac{dp_0(t)}{dt} = -\lambda p_0(t)$$

which can also be written as

$$[\log p_0(t)]' = -\lambda$$

By antidifferentiation, $p_0(t) = e^{-\lambda t + c}$, where c is an arbitrary constant. However, since in (3.41), the numerator must converge to zero, $p_0(0) = 1$, and thus $c = 0$; consequently, we established the theorem for $k = 0$. If we now utilize the known form of $p_0(t)$ in (3.44), and let $\triangle t \to 0$, we get from (3.40), (3.41), and (3.44),

$$[p_1(t)]' = -\lambda p_1(t) + \lambda e^{-\lambda t}$$

If we multiply this equation by $e^{\lambda t}$, it becomes

$$[e^{\lambda t} p_1(t)]' = \lambda$$

from which, by antidifferentiaiton, $e^{\lambda t} p_1(t) = \lambda t + d$, where d is an arbitrary constant. But because from (3.40) $p_1(0) = 0$, $d = 0$, and thus the theorem is established for $k = 1$ as well. Now, proving the theorem by induction, we assume that it has been proved for $k - 1$, which we apply in (3.45); if we let $\triangle t \to 0$ again, (3.40), (3.41), (3.45), (3.42), and condition (iv) yield

$$[p_k(t)]' = -\lambda p_k(t) + \lambda \frac{(\lambda t)^{k-1} e^{-\lambda t}}{(k-1)!}$$

which can be solved by the same trick as applied in the case of $k = 1$, yielding the desired form for $p_k(t)$. This completes the proof.   ▲

We postpone the discussion of Theorem 3.10 through examples to the next chapter, where the introduction of the concept of expectation enables us to make the examples more interesting and more meaningful.

3.12 EXERCISES

1. Let us choose $\Omega = (0, 1)$ for the experiment of tossing a coin with the dictionary of $H = (0, 1/2)$ and $T = [1/2, 1)$. List all events in this experiment (as subsets of Ω), and conclude that the function $f(x) = x$ for $0 < x < 1$ is not a random variable. Is the function $g(x) = 1/2$ if $0 < x < 1/2$, and $g(x) = -2$ if $1/2 \le x < 1$, a random variable?

2. Assume the following distribution function of the random variable X:

$$F(x) = \begin{cases} 0 & \text{if } x < 0 \\ x & \text{if } 0 \leq x < \dfrac{1}{2} \\ A(x-1)^2 + B & \text{if } \dfrac{1}{2} \leq x < 1 \\ 1 & \text{if } x \geq 1 \end{cases}$$

Determine A and B (i) if $F(x)$ is known to be continuous; and (ii) if $P(X = 1/2) = 1/4$ and $P(X = a) = 0$ for all $a \neq 1/2$.

3. The functions (i) $F(x) = e^{-|x|}$ for all x; (ii) $F(x) = 1 - 1/x^2$ if $x \geq 0.8$ and $F(x) = 0$ if $x < 0.8$; and (iii) $F(x) = 0$ if $x < 3$ and $F(x) = (x-2)(3x+1)/(x^2-4)$ if $x \geq 3$ cannot be the distribution function of a random variable. Why?

4. Why is there a contradiction between the assumptions of $P(X = a) = 0$ for all a, and the distribution function of X is $F(x) = 0$ if $x < 2$ and $F(x) = 1 - 1/x^2$ if $x \geq 2$?

5. The distribution function of the random variable X,

$$F(x) = \begin{cases} 0 & \text{if } x < -1 \\ x + 1.5 & \text{if } -1 \leq x < -\dfrac{1}{2} \\ 1 & \text{if } x \geq -\dfrac{1}{2} \end{cases}$$

Find (i) $P(X = -2)$, $P(X = -1)$, $P(X = -1/2)$, and $P(X = 1)$; (ii) $P(-1 < X < 0)$; (iii) $P(-0.2 \leq X \leq -0.4)$; and (iv) $P(-0.2 < X < -0.4)$.

6. By repeating the argument of Section 3.3, show that if X is a positive integer-valued random variable such that for all nonnegative integers t and s,

$$P(X > t + s \mid X > t) = P(X > s)$$

then $P(X = s) = p(1-p)^{s-1}$, $s \geq 1$ (the geometric distribution).

7. The possible values of the (discrete) random variable X are -2, 1, 3, and 4, whose distribution satisfies

$$P(X = -2) = P(X = 1) = 2P(X = 3) = 3P(X = 4)$$

Determine the value $P(X = a)$ for all (possible) a.

8. Let $F(z)$ be the distribution function of the random variable X of Exercise 7. Find $F(-2.1)$, $F(0.8)$, $F(3.2)$, and $F(4)$. In addition, evaluate $P(-1.8 < X \leq 3)$.

9. State if the random variable X whose distribution function

$$F(z) = \begin{cases} 0 & \text{if } z < -1 \\ \dfrac{1}{4} & \text{if } -1 \leq z < 2 \\ \dfrac{3}{4} & \text{if } 2 \leq z < 5 \\ 1 & \text{if } z \geq 5 \end{cases}$$

is discrete. If your answer is yes, give the values and the distribution of X.

10. State if the random variable X whose distribution function,

$$F(z) = \begin{cases} 0 & \text{if } z < -1 \\ z + 1 & \text{if } -1 \le z < -\dfrac{1}{2} \\ 1 & \text{if } z \ge -\dfrac{1}{2} \end{cases}$$

is discrete. If your answer is yes, give the values and the distribution of X.

11. Urn 1 contains three red and two white balls, while urn 2 is composed of two red and two white balls. A coin if flipped, and if it lands heads, two balls are selected at random with replacement from urn 1, whereas in the case of tails, two balls are selected at random without replacement from urn 2. Let X be the number of red balls among those selected. Give the distribution of X.

12. Let X have a hypergeometric distribution with parameters $T = 20$, $M = 8$, and $t = 6$. Find $P(X = 5 \mid X \ge 2)$.

13. Let X have a binomial distribution with parameters $n = 20$ and $p = 0.4$. Find $P(X = 5 \mid X \ge 2)$.

14. There are a random number X of eggs on a farm, whose distribution is assumed to be Poisson (with parameter λ). A team of raccoons raids the farm, and they take every egg with probability 0.02. Assuming that the events whether different eggs are taken or not are independent, determine the distribution of the number of eggs left on the farm.

15. Find the distribution of the number of times a fair die has to be rolled to get a 6.

16. A pair of fair dice is rolled as long as the sum of the two faces is different from 7. What is the probability that the experiment terminates in fewer than eight rolls? Compare the general distribution of the number of rolls required in this experiment with that of Exercise 15.

17. What is the distribution of the indicator variable of the event that in 10 rolls of a regular die, the number 6 will turn up exactly twice?

18. For each of the following distribution functions $F(z)$, give the density function: (i) $F(z) = 1 - 1/z^2$, $z \ge 1$; (ii) $F(z) = [1/(2\pi)] \int_0^z \sin^2 x \, dx$, $0 < z < 4\pi$; and (iii) $F(z) = (2/\pi)$ arc sin $\sqrt{z}$, $0 < z < 1$. In each case, give the value of the density function at $z = -1$.

19. The function $f(z) = Cz^{-5}$, $z \ge 1$, and $f(z) = 0$ for $z < 1$, is a density function. Determine C, and give the corresponding distribution function.

20. The random variable X is uniformly distributed over the interval $(-2, 5)$. Evaluate (i) the distribution function of X; and (ii) $P(X < 2 \mid X > 0)$.

21. The random variable X is normal with parameters $m = -2$ and $\sigma = 5$. We then know that $Y = 3 - 5X$ is also normal. Find its parameters. Use Table 2.1 to find $P(10 < Y < 20)$.

22. The random variable Y is Poisson with parameter $\lambda = 0.8$, and X is another random variable, whose conditional distribution is binomial with parameters k

and $p = 0.3$, given that $Y = k$. Find (i) $P(X = 2, Y = 3)$; and (ii) the distribution of X.

23. The joint distribution of X and Y is given in the following table:

X \ Y	1	2	3	4
−1	0.1	0.1	0	0.3
2	0.2	0	0.1	0.1
3	0	0.1	0	0

Find (i) the distribution of X; (ii) the distribution of Y; (iii) $P(Y \le 2 \mid X = 2)$; and (iv) $P(X + Y = 3)$.

24. From an ordinary deck of 52 cards, 5 are dealt out. Let X and Y, respectively, be the number of aces and kings among those dealt. Determine the joint distribution of X and Y.

25. Let X be a random variable with distribution function $F(x)$. Show that the bivariate distribution function of the vector (X, X) is $F(x, y) = \min (F(x), F(y))$. What are the univariate marginals of $F(x, y)$? Deduce from this example that a bivariate distribution function whose univariate marginal distributions are normal need not be bivariate normal.

26. The density function of the vector (X, Y),

$$f(x, y) = \begin{cases} Cxy & \text{if } 0 < y < |x|, \quad -2 < x < 2 \\ 0 & \text{otherwise} \end{cases}$$

Determine (i) the value of C; and (ii) the marginal distribution functions and density functions. In addition, evaluate $P(-1 < X < 1, 1/2 < Y < 1)$.

27. For the vector of Exercise 26, find the conditional density of X, given $Y = y$. Hence compute $P(X < 1 \mid Y = 1/2)$.

28. Given $Y = y$, the density function of X is exponential; that is, $f(x \mid y) = ye^{-yx}$, $x > 0$. Find the bivariate density of (X, Y) if Y itself is exponential with density function $f_Y(y) = e^{-y}$, $y > 0$.

29. Are the components of (X, Y) of Exercise 26 independent?

30. The random variables X and Y are independent normal variables with parameters $m_1 = -1$, $\sigma_1 = 2$ and $m_2 = 3$, $\sigma_2 = 5$, respectively. Use Table 2.1 to calculate $P(X \ge 2, 0 < Y < 12)$.

31. Let X be uniformly distributed on $(0, 1)$, and let the distribution function of Y be $F(y) = 1 - 1/y^2$, $y \ge 1$. Find the probability $P(XY \le 5)$ if X and Y are independent. Generalize to $P(XY \le z)$.

32. The random variables X, Y, and Z are independent unit exponential variables; that is, their common distribution function is $F(x) = 1 - e^{-x}$, $x > 0$. Find the probability that (i) the largest is smaller than 2; and (ii) the second largest is larger than 1.

33. The random variables X and Y are independent exponential variables with the same parameter. Find $P(X < 2Y)$.

34. Let X and Y be independent random variables, where X is exponential and Y is uniform on $(0, 1)$. Find the density function of $X + Y$.

35. The random variables X, Y, and Z are independent and uniformly distributed on $(0, 1)$. Find the density function of $X + Y$ and $X + Y + Z$.

36. The random variables X and Y are independent discrete variables with $P(X = 1) = P(X = 2) = P(X = 3) = 1/3$, and $P(Y = 0) = 0.1$, $P(Y = 1) = 0.5$, and $P(Y = 3) = 0.4$. Find the distribution of $X + Y$.

37. Let X and Y be independent Poisson variables with the same parameter. Find $P(Y = y \mid X + Y = z)$.

38. Show that if the random variable X satisfies $0 < X < 1$, and if $P(X \le uv \mid X \le u) = P(X \le v)$ for all $0 < u$, $v < 1$, then X is uniformly distributed.

39. Let X be a standard normal variable and define $Y = 10(X^{0.1} - 1)$. Compute $P(Y \le 0)$, $P(Y \le 0.2)$, $P(Y \le 0.4)$, and $P(Y \le 1)$, and compare these values with $P(X \le u)$, where $u = 0, 0.2, 0.4$, and 1, taken from Table 2.1. Finally, compare $P^{365}(Y \le u)$ and $P^{365}(X \le u)$ for the previous four values of u. Comment on these values in terms of the flood problem in Section 3.11.

40. Translate the assumptions of a Poisson process into terms of the following practical problems: (i) on an imaginary line of the sky, the stars form a Poisson process; (ii) the points of time when a car enters a turnpike at a fixed entry form a Poisson process, (iii) the points of time when the atoms of a radioactive material split form a Poisson process, and (iv) the points of time of earthquakes around city L form a Poisson process.

4
Expectation and Variance

4.1 THE EXPECTED VALUE OF DISCRETE RANDOM VARIABLE

Assume that the profit X on an item of a product line is random, which is -2, 3, or 5, depending on whether the item is defective (cost of production is lost), sold at a discount, or sold at "full price." Assume further that past experience shows that $p_1 = P(X = -2) = 0.05$, $p_2 = P(X = 3) = 0.35$, and $p_3 = P(X = 5) = 0.6$. The company's interest evidently is

$$(4.1) \qquad Y_n = X_1 + X_2 + \cdots + X_n$$

where n is the number of items manufactured in a quarter, say, and X_j is the profit on the jth item. A remarkable result of probability theory, which we deduce here from the relation of the relative frequency and the probability of an event (see Section 2.6), is that the aggregate Y_n, for large n, has so little relative random fluctuation that it can be substituted by the constant nE, where E is a number computable from the distribution of X. The company can, therefore, plan ahead as to its quarterly profits in the future (as long as the economic conditions keep the distribution of X unchanged; of course, if management detects a change in the economy, it can plan with the same accuracy by utilizing a new distribution of X, appropriate for the changing economy).

The claim on Y_n is proved as follows. Every term of Y_n is either -2, or 3, or 5, and thus by grouping the (-2)s, the 3s, and the 5s together, we get

$$(4.2) \qquad Y_n = -2k_1(n) + 3k_2(n) + 5k_3(n)$$

where $k_1(n)$, $k_2(n)$, and $k_3(n)$ are the numbers of the (-2)s, the 3s, and the

5s, respectively, among the X_j, $1 \leq j \leq n$. Now, we assume the X_j, $1 \leq j \leq n$, to be independent, and each X_j is distributed as X (each is a copy of X). This means, among other things, that the events $\{X_j = -2\}$, $1 \leq j \leq n$, are independent, and $P(X_j = -2) = P(X = -2) = 0.05$. Therefore, $k_1(n)$, in fact, is the frequency of $A_1 = \{X = -2\}$ in n independent repetitions of the experiment concerning A_1 (i.e., produce and sell the products in question). We can argue similarly to conclude that $k_2(n)$ is the frequency of $A_2 = \{X = 3\}$, and $k_3(n)$ is the frequency of $A_3 = \{X = 5\}$ in n independent repetitions of the same experiment. Consequently, by the relation of the relative frequency and the probability of an event,

$$k_1(n) \sim np_1 = 0.05n; \qquad k_2(n) \sim np_2 = 0.35n; \qquad k_3(n) \sim np_3 = 0.6n$$

and thus, from (4.2),

$$Y_n \sim -2np_1 + 3np_2 + 5np_3 = n(-0.1 + 1.05 + 3) = 3.95n$$

In this argument, the values $x_1 = -2$, $x_2 = 3$, and $x_3 = 5$ had no specific roles, and neither had the actual values of p_1, p_2, and p_3. In other words, we can repeat the argument above without any change to prove the following result.

Theorem 4.1 Let X be a discrete random variable whose values are $x_1, x_2, \ldots, x_s$, and whose distribution is $p_j = P(X = x_j)$, $1 \leq j \leq s$. Let $X_1, X_2, \ldots, X_n$ be independent copies of X. Then, for large n,

$$(4.3) \qquad Y_n = X_1 + X_2 + \cdots + X_n \sim nE$$

where

$$(4.4) \qquad E = x_1 p_1 + x_2 p_2 + \cdots + x_s p_s$$

An equivalent form of (4.3) is that the arithmetical mean

$$(4.5) \qquad M_n = \frac{Y_n}{n} = \frac{X_1 + X_2 + \cdots + X_n}{n} \sim E$$

Although we could utilize the results of Section 2.6 in estimating the error terms of the approximation in (4.3) or (4.5), we postpone the discussion of error terms to Section 4.5, which deals specifically with such error estimation.

Because of the significance of the number E introduced in (4.4), we assign a name to it.

Definition 4.1 Let X be a discrete random variable with values $x_1, x_2, \ldots, x_s$, and with distribution $p_j = P(X = x_j)$, $1 \leq j \leq s$. Then the number $E = E(X)$ of (4.4) is called the *expectation*, or the *expected value*, of X.

Example 4.1 Let A be an event, and let $I(A)$ be its indicator variable; that is, $I(A) = 1$ if A occurs, and $I(A) = 0$ if A fails to occur. Then $E[I(A)] = P(A)$.

As a matter of fact, from (4.4),

$$E[I(A)] = 1 \times P(A) + 0 \times [1 - P(A)] = P(A) \qquad \blacktriangle$$

Example 4.2 Roll a regular die and let X be the number that comes up on it. Let us find $E(X)$.

With the notations of Definition 4.1, $x_j = j$, $1 \le j \le 6$, and $p_j = 1/6$ for each j. Hence, by (4.4),

$$E(X) = \frac{1}{6}(1 + 2 + 3 + 4 + 5 + 6) = \frac{21}{6} = 3.5 \qquad \blacktriangle$$

We now see that the expected value of X does not have to be one of the values of X. The significance of $E(X)$ is in its relation to the average M_n of (4.5), not directly to X itself.

Example 4.3 Let the distribution of X be binomial with parameters n and p. Then $E(X) = np$.

We have actually computed $E(X)$ in the Lemma of Section 2.8. Indeed, with the notations of Definition 4.1, $x_{m+1} = m$, $0 \le m \le n$, and $p_{m+1} = \binom{n}{m}p^m(1 - p)^{n-m}$. Hence the definition at (4.4) becomes formula (2.29). $\qquad \blacktriangle$

Using an argument different from the direct proof of Theorem 4.1, we shall later show that (4.3), or equivalently (4.5), remains valid for discrete random variables taking on infinitely many values, in which case E is to be computed as given below.

Definition 4.2 Let X be a discrete random variable taking on the values x_j, $1 \le j < +\infty$, with distribution $p_j = P(X = x_j)$. We call

$$(4.6) \qquad\qquad E = E(X) = \sum_{j=1}^{+\infty} x_j p_j$$

the expected value, or expectation, of X, whenever the infinite series in (4.6) is *absolutely convergent*. When this infinite series is not absolutely convergent, we say that the expected value of X does not exist, or that X does not have a finite expectation.

The reason conditional convergence (i.e., not absolute convergence) is not sufficient is the fact, well known in calculus, that if the values x_j (and thus the corresponding p_j as well) are listed in a different order, the infinite sum in (4.6) will change its value, while we still speak of the same random variable X.

Example 4.4 Let X be a geometric variable with parameter p. Then $E(X) = 1/p$.

By definition, X takes on the positive integers with $p_j = P(X = j) = p(1 - p)^{j-1}$. Hence, by Definition 4.2,

$$(4.7) \qquad E(X) = \sum_{j=1}^{+\infty} jp(1 - p)^{j-1} = p\sum_{j=1}^{+\infty} j(1 - p)^{j-1}$$

if it is convergent (since every term is positive, it is then absolutely convergent). Now, if we differentiate the sum [see (1.25)]

$$\sum_{j=0}^{+\infty} x^j = \frac{1}{1 - x} \qquad |x| < 1$$

we get

$$\sum_{j=1}^{+\infty} jx^{j-1} = \frac{1}{(1 - x)^2}$$

which is exactly the sum needed in (4.7) with $x = 1 - p$. We thus have

$$E(X) = p\frac{1}{[1 - (1 - p)]^2} = \frac{1}{p} \qquad\qquad \blacktriangle$$

Example 4.5 Let us use asymptotic results to compare the answers to the following questions: (i) how many times do we have to toss a fair coin to get a head 100 times; and (ii) how many times does a head turn up if a fair coin is tossed 200 times?

When answering question (i), we assume once again that (4.3) is valid with E being computed by (4.6). Therefore, the requirement of getting exactly 100 heads means that we repeat, independently of each other, $n = 100$ times the experiment of waiting for a head. That is, if X is the random variable equal to the number of times a fair coin has to be tossed in order to get a head, then the answer to (i) is $Y_{100} = X_1 + X_2 + \cdots + X_{100}$, where the X_j are independent copies of X. Now, the distribution of the discrete waiting time is geometric, whose parameter, in our particular case, $p = 1/2$. Hence, by Example 4.4 and by (4.3), $Y_{100} \sim 100E(X) = 200$.

On the other hand, when a fair coin is tossed 200 times, the frequency k of heads satisfies the asymptotic relation $k \sim 200p = 100$. $\qquad\qquad \blacktriangle$

A similar comparison can be repeated for an arbitrary event, whose probability is p, from which it can be seen that asymptotically, we get the same result whether we want a specific frequency in a random number of repetitions, or when we specify the number of repetitions of the experiment in advance and we count the (random) frequency.

Example 4.6   Let the distribution of X be Poisson with parameter λ. Then $E(X) = \lambda$.

For a Poisson variable X, the values x_j are the nonnegative integers, and its distribution $p_j = P(X = j) = \lambda^j e^{-\lambda}/j!$. Therefore, by (4.6) [we start with $j = 1$ in the summation below, because the contribution of $j = 0$ is 0]

$$E(X) = \sum_{j=1}^{+\infty} j \frac{\lambda^j e^{-\lambda}}{j!} = \lambda e^{-\lambda} \sum_{j=1}^{+\infty} \frac{\lambda^{j-1}}{(j-1)!} = \lambda e^{-\lambda} e^{\lambda} = \lambda$$

where, in the last sum, we used the Taylor formula for e^{λ}.   ▲

Example 4.7   Assume that telephone calls arrive at a switchboard according to a Poisson process. If the expected number of calls in a 5-minute period is 10, what is the probability that in a 1-minute period exactly one call is received?

Recall Definition 3.9 and Theorem 3.10 concerning the Poisson process. In view of the Poissonian character of the number $X(t)$ of calls in a time period of length t,

$$(4.8) \qquad\qquad P(X(t) = k) = \frac{(\lambda t)^k e^{-\lambda t}}{k!}$$

Now, if time is measured in minutes, the assumption is that $E(X(5)) = 10$. On the other hand, from (4.8) and from Example 4.6, $E(X(5)) = 5\lambda$. These two forms yield $\lambda = 2$. Therefore, by one more appeal to (4.8),

$$P(X(1) = 1) = \lambda e^{-\lambda} = 2e^{-2} = 0.27 \qquad\qquad ▲$$

We now return to our theoretical discussion and establish the basic properties of expectation.

Theorem 4.2   For discrete random variables X and Y, and for arbitrary constants a and b,
(i) $E(aX + b) = aE(X) + b$,
(ii) $E(X + Y) = E(X) + E(Y)$,
whenever the right-hand sides are meaningful. Furthermore, if X and Y are independent, and both $E(X)$ and $E(Y)$ are defined,
(iii) $E(XY) = E(X)E(Y)$.

Remark   It follows by induction that (ii) extends to sums with a finite number of terms. A similar extension also applies to (iii).

Proof: In all summations in this proof, Σ is a finite or an infinite sum according as the indicated subscript constitutes a finite or infinite set.

(i)   Let the values of X be $x_1, x_2, \ldots$. Then the values of $aX + b$ are $ax_1 + b$, $ax_2 + b, \ldots$, but the distribution of X and $aX + b$ is the same sequence $p_1, p_2, \ldots$. Hence (summations are over j)

$$E(aX + b) = \sum (ax_j + b)p_j = \sum (ax_j p_j + bp_j) = a\sum x_j p_j + b\sum p_j$$

where $\sum p_j = 1$ (for any distribution) and $\sum x_j p_j = E(X)$.

(ii)   In addition to the notations in (i), we introduce $y_1, y_2, \ldots$ for the values of Y, $r_1, r_2, \ldots$ for the distribution of Y, and $p(i, j) = P(X = x_i, Y = y_j)$. Now, since (see Section 3.7)

$$\sum_i p(i, j) = r_j, \qquad \sum_j p(i, j) = p_i$$

$$E(X) = \sum_i x_i p_i = \sum_i x_i \left(\sum_j p(i, j) \right) = \sum_i \sum_j x_i p(i, j)$$

and

$$E(Y) = \sum_j y_j r_j = \sum_j y_j \left(\sum_i p(i, j) \right) = \sum_i \sum_j y_j p(i, j)$$

where the change of the order of summation is permitted in view of the absolute convergence of the sums. Consequently,

$$E(X) + E(Y) = \sum_i \sum_j (x_i + y_j)p(i, j) = \sum_i \sum_j (x_i + y_j)P(X = x_i, Y = y_j)$$

which is indeed $E(X + Y)$; namely, if we collect all terms i and j such that $x_i + y_j = s$, the sum of $P(X = x_i, Y = y_j)$ over these i and j yields $P(X + Y = s)$.

(iii) If X and Y are independent, then $p(i, j) = p_i r_j$, and

$$E(X)E(Y) = \left(\sum_i x_i p_i \right)\left(\sum_j y_j r_j \right) = \sum_i \sum_j x_i y_j p_i r_j$$

$$= \sum_i \sum_j x_i y_j p(i, j) = \sum_i \sum_j x_i y_j P(X = x_i, Y = y_j)$$

Arguing as at the end of part (ii), the last sum is $E(XY)$, because if we collect those terms i and j for which $x_i y_j = s$, and sum over these i and j first, the sum of $P(X = x_i, Y = y_j)$ becomes $P(XY = s)$. Hence the sum over all i and j above becomes the sum of $sP(XY = s)$ over s, which is $E(XY)$.

The proof is completed.   ▲

Example 4.8   For the (discrete) random variables X and Y, $E(X) = 2$ and $E(Y) = -3$. Let us find $E(3X - 5Y)$.

By part (ii) of Theorem 4.2, $E(3X - 5Y) = E(3X) + E(-5Y)$, which, by part (i), yields

$$E(3X - 5Y) = 3E(X) - 5E(Y) = 6 + 15 = 21 \qquad \blacktriangle$$

Example 4.9   Let X and Y be independent Poisson variables with $E(X) = 2$ and $E(3X - 2XY) = 4$. Let us find $P(Y = 0)$.

By Theorem 4.2 and by the assumptions,

$$4 = E(3X - 2XY) = E(3X) + E(-2XY) = 3E(X) - 2E(X)E(Y) = 6 - 4E(Y)$$

so $E(Y) = 1/2$. Now, for a Poisson variable, the expectation is its parameter; thus $\lambda = 1/2$ for Y. Consequently, $P(Y = 0) = e^{-1/2} = 0.607$. $\blacktriangle$

The utilization of decomposing random variables as sums of indicator variables is particularly advantageous for computing expectation.

Example 4.10   A new way of computing $E(X)$ for a binomial variable with parameters n and p is as follows. Recall that X is the frequency of a specific event A with $P(A) = p$ in n independent repetitions of the experiment defining A. Let

$$I_j = \begin{cases} 1 & \text{if } A \text{ occurs at the } j\text{th repetition} \\ 0 & \text{otherwise} \end{cases}$$

Then $X = I_1 + I_2 + \cdots + I_n$. We thus get $E(X) = np$.

Indeed, since for each j, $E(I_j) = p$, Theorem 4.2 yields $E(X) = E(I_1) + E(I_2) + \cdots + E(I_n) = np$. $\blacktriangle$

Example 4.11   Let X be a hypergeometric variable; that is, let a lot contain M type I and $T - M$ type II items. We select t items at random and without replacement from the lot, and X is the number of type I items among those selected. Then $E(X) = tM/T$ and $E(X(X - 1)) = t(t - 1)M(M - 1)/T(T - 1)$.

Namely, introducing

$$I_j = \begin{cases} 1 & \text{if the } j\text{th item selected is type I} \\ 0 & \text{otherwise} \end{cases}$$

we have $X = I_1 + I_2 + \cdots + I_t$. The reader was asked in Exercise 31 of Chapter 1 to show (in terms of probabilities) that $E(I_j) = M/T$ for each j. Hence, by Theorem 4.2,

$$E(X) = E(I_1) + E(I_2) + \cdots + E(I_t) = \frac{tM}{T}$$

By the same argument as above,

$$E[X(X-1)] = E(X^2) - E(X) = E[(I_1 + I_2 + \cdots + I_t)^2] - E(I_1 + I_2 + \cdots + I_t)$$

$$= E(I_1^2 + I_2^2 + \cdots + I_t^2) + 2 \sum_{1 \le j < k \le t} E(I_j I_k) - E(I_1 + I_2 + \cdots + I_t)$$

Note that $I_j^2 = I_j$, and thus

$$E(I_1^2 + I_2^2 + \cdots + I_t^2) - E(I_1 + I_2 + \cdots + I_t) = 0$$

On the other hand, from the exercise cited,

$$E(I_j I_k) = \frac{M(M-1)}{T(T-1)} \qquad \text{for all } 1 \le j < k \le t$$

The combination of these yields

$$E[X(X-1)] = 2\binom{t}{2}\frac{M(M-1)}{T(T-1)}$$

as claimed.   ▲

Example 4.12   Assume that a cereal company puts into each cereal box one of three different cards, and a person, who has one of each of the three cards, gets a refund. The company ensures that the same number of each type of card is placed into its products (thus, getting a specific type of card in a box has probability 1/3), and the contents of the boxes concerning cards are independent events. What is the expected number of boxes that one has to purchase in order to get the refund?

The first purchase provides the first card. Then let X be the number of boxes to be purchased in order to get a second type of card, and after that, Y is the number of boxes required for getting the third type of card. The desired value is

$$E(1 + X + Y) = 1 + E(X) + E(Y)$$

where Theorem 4.2 was applied. Now, both X and Y are geometric variables, because they represent random waiting times. The parameter of X is $p = 2/3$ (two out of three cards are favorable), and the parameter of Y is $p^* = 1/3$ (only one kind is favorable at this stage). Example 4.4 thus yields $1 + 1/p + 1/p^* = 1 + 3/2 + 3 = 5.5$ for the desired expected value.   ▲

Note the value $E(Y) = 3$, "suggesting" that a certain type of card is more difficult to get than the others. The contrast is even larger if more than three types of card are to be collected. Another aspect of the example to be noted is the fact that, by Theorem 4.1, the company knows in advance the cost of the offer. The company can therefore offset this cost either by including it in the price of the cereal, or by charging it against advertisement expenditures (which will sooner or later be a part of the price as well).

4.2   THE EXPECTATION OF ABSOLUTELY CONTINUOUS RANDOM VARIABLE

Let X be a random variable with density function $f(x)$. Let $X_1, X_2, ..., X_n$ be independent copies of X, and as in the discrete case, set

$$M_n = \frac{X_1 + X_2 + \cdots + X_n}{n}$$

Our aim again is to show that there is a number E such that, for large n, in some sense, M_n is asymptotically E. In Section 4.5 we shall establish this property of M_n, in which E is to be computed by the formula below.

Definition 4.3   The expected value, or expectation, of a random variable X with density function $f(x)$ is defined by the formula

$$(4.9) \qquad E = E(X) = \int_{-\infty}^{+\infty} xf(x) \, dx$$

whenever this (improper) integral is absolutely convergent. If the integral in (4.9) is not absolutely convergent, we say that X does not have an expected value (or that its expectation is not finite).

Example 4.13   Let X be an exponentially distributed random variable; that is, its density function,

$$f(x) = \begin{cases} \lambda e^{-\lambda x} & \text{if } x \geq 0 \\ 0 & \text{otherwise} \end{cases}$$

where $\lambda > 0$. Then $E(X) = 1/\lambda$.

We apply (4.9) and integrate by parts, getting

$$E(X) = \int_{-0}^{+\infty} \lambda x e^{-\lambda x} \, dx = -x e^{-\lambda x} \Big]_0^{+\infty} + \int_0^{+\infty} e^{-\lambda x} \, dx$$

where the first term is zero and the integral in the second term is indeed $1/\lambda$.

▲

Example 4.14   Assume that the random life X of an electric light bulb is exponentially distributed with $E(X) = 1000$ hours. Let us evaluate $P(X \leq 500)$.

By the assumption, the distribution function of X, $F(x) = 1 -$

$e^{-\lambda x}$, $x \geq 0$, where, by Example 4.13, $\lambda = 0.001$. Hence the desired probability

$$P(X \leq 500) = F(500) = 1 - e^{-0.001 \times 500} = 1 - e^{-0.5} = 0.393 \qquad \blacktriangle$$

Example 4.15   Assume again that the random life length X of an electric light bulb is exponentially distributed with $E(X) = 1000$ hours. If we buy 100 bulbs, each with the characteristics of X, how soon can we expect the first one to burn out?

Let us label the bulbs using the numbers 1 through 100, and let X_j be the life length of the jth bulb. Then the bulb with the shortest (random) life will last

$$T = \min(X_1, X_2, \ldots, X_{100})$$

hours. Assuming that the X_j are independent, and utilizing the fact that each X_j is distributed as X, for $x > 0$

$$P(T > x) = P(X_1 > x, X_2 > x, \ldots, X_{100} > x)$$

$$= P(X_1 > x)P(X_2 > x) \cdots P(X_{100} > x) = (e^{-0.001x})^{100} = e^{-0.1x}$$

that is, the distribution function $P(T \leq x) = 1 - e^{-0.1x}$ of T is exponential, again, with parameter 0.1. Hence, by Example 4.13, $E(T) = 10$. $\blacktriangle$

Although the result of Example 4.15 is surprising (even somewhat shocking for the unsuspecting customer), it does not contradict our earlier claim regarding the average M_n. That is, out of many exponentially distributed bulbs, or other products, one might lose with the one that burns out first, but the longest-lasting bulb will compensate for it. One more surprise to think about in this connection: Assume that the bulb which has just burned out is replaced immediately by a new one. Is it possible that the very next one to burn out is this new bulb? The answer is yes, due to the no-aging property (which property is also known by the name "lack of memory") of the exponential distribution. Remember that the no-aging property means that the origin of the time scale can be chosen anywhere. So we can count time from the moment a bulb is replaced, and thus all bulbs act equally.

Example 4.16   Assume once more that the random life length X of an electric bulb is exponential with $E(X) = 600$ hours. The manufacturer, however, cheats and claims that the bulbs last an average of 1000 hours. To back up this claim, a complete refund is given if the bulb lasts less than 800

hours. At which price would this manufacturer break even if the cost of
production is 20 cents per bulb?

Let the price of the bulb be p, and let the profit on a bulb be Y. By the
policy noted above,

$$Y = \begin{cases} p - 20 & \text{if } X \geq 800 \\ -20 & \text{if } X < 800 \end{cases}$$

The manufacturer's interest is the average profit on a large number of bulbs,
which, by Theorem 4.1, approximately equals $E(Y)$. Therefore, we have to
determine p such that $E(Y) = 0$. By definition, and by Example 4.13,

$$E(Y) = (p - 20)P(X \geq 800) - 20P(X < 800)$$
$$= (p - 20)e^{-800/600} - 20(1 - e^{-800/600}) = pe^{-4/3} - 20$$

from which $E(Y) = 0$ yields $p = 20e^{4/3} = 75.87$ cents.   ▲

Although the numbers in the example might be unrealistic, its aim is to
bring out sharply the point that a warranty that sounds very good might be
the major contribution to the price of the product.

Example 4.17 Assume that customers arrive at a store according to a
Poisson process. What is the expected number of customers in a 2-hour
period if the expected time to the arrival of the first customer is 2 minutes?

Let us measure time in minutes, so we translate 2 hours to 120
(minutes). Now, by Theorem 3.10 and the Corollary to it, the waiting time
for the first customer to arrive is exponential in a Poisson process. Thus,
from Example 4.13, its parameter λ satisfies $1/\lambda = 2$, (i.e., $\lambda = 1/2$). With this
same λ, however, the number of customers arriving in a 120-minute period
has Poisson distribution with parameter $120\lambda = 60$. Consequently, the
desired expectation is 60 (in view of Example 4.6).   ▲

The expectation of absolutely continuous random variables has the
same basic properties as those established in the discrete case.

Theorem 4.3 For absolutely continuous random variables X and Y, and for
arbitrary constants $a \neq 0$ and b,
(i) $E(aX + b) = aE(X) + b$,
(ii) $E(X + Y) = E(X) + E(Y)$,
whenever the right-hand sides have been defined. Furthermore, if X and Y
are independent, and both $E(X)$ and $E(Y)$ are defined,
(iii) $E(XY) = E(X)E(Y)$.

Remark   Once again, just as in the discrete case, (ii) and (iii) can be extended by induction to an arbitrary number of terms.

Proof: (i) Let the density function of X be $f(x)$. We first establish that the density function of $(aX + b)$ is $|1/a|f[(x - b)/a]$. That is, if $a > 0$,

$$P(aX + b \leq x) = P\left(X \leq \frac{x - b}{a}\right) = F\left(\frac{x - b}{a}\right)$$

from which differentiation yields the claimed form of the density. Now, if $a < 0$,

$$P(aX + b \leq x) = P\left(X \geq \frac{x - b}{a}\right) = 1 - F\left(\frac{x - b}{a}\right)$$

whose derivative is $-(1/a)f((x - b)/a)$, which again is the claimed form. Therefore, by (4.9) and the substitution $y = (x - b)/a$,

$$E(aX + b) = \int_{-\infty}^{+\infty} x \frac{1}{|a|} f\left(\frac{x - b}{a}\right) dx = \int_{-\infty}^{+\infty} (ay + b)f(y) \, dy$$

$$= a \int_{-\infty}^{+\infty} yf(y) \, dy + b \int_{-\infty}^{+\infty} f(y) \, dy = aE(X) + b$$

where once more, (4.9) and a general property of density were utilized.

(ii) Let $f(x, y)$ be the density function of the vector (X, Y), and let the density function of Y be $g(y)$. Then, by (3.23),

$$(4.10) \quad P(X + Y \leq z) = \iint_{x+y \leq z} f(x, y) \, dx \, dy = \int_{-\infty}^{+\infty} \left[\int_{-\infty}^{z-y} f(x, y) \, dx\right] dy$$

and by (3.26) and (3.27),

$$(4.11) \qquad \int_{-\infty}^{+\infty} f(x, y) \, dx = g(y), \qquad \int_{-\infty}^{+\infty} f(x, y) \, dy = f(x)$$

Differentiation in (4.10) gives the density function of $X + Y$:

$$h(z) = \int_{-\infty}^{+\infty} f(z - y, y) \, dy$$

Therefore, from (4.9),

$$E(X + Y) = \int_{-\infty}^{+\infty} zh(z) \, dz = \int_{-\infty}^{+\infty} z\left[\int_{-\infty}^{+\infty} f(z - y, y) \, dy\right] dz$$

which, with the substitution $u = z - y$, becomes

$$E(X + Y) = \int_{-\infty}^{+\infty} (u + y) \int_{-\infty}^{+\infty} f(u, y) \, dy \, du$$

$$= \int_{-\infty}^{+\infty} u \int_{-\infty}^{+\infty} f(u, y) \, dy \, du + \int_{-\infty}^{+\infty} y \int_{-\infty}^{+\infty} f(u, y) \, du \, dy$$

where the interchange of the order of integration in the second term is justified by the absolute convergence of the (improper) integrals involved. By an appeal to (4.11) we now get that the right-hand side is $E(X) + E(Y)$.

(iii) The proof of this part is very similar to that of part (ii). The formula in (4.10) is to be replaced by integration over $xy \leq z$, and the independence of X and Y is to be utilized through the equation $f(x, y) = f(x)g(y)$ [which makes (4.11) superfluous in this case]. After getting the density $h^*(z)$ of XY, $E(XY)$ is computed by (4.9), which can be recognized to split into the product $E(X)E(Y)$ after a simple substitution. Further details are omitted.

The theorem is established.   ▲

Example 4.18 Let X be normally distributed with parameters m and σ. Then $E(X) = m$.

We utilize the representation $X = \sigma U + m$, where the distribution of U is standard normal [see (3.13)]. We thus show that $E(U) = 0$, which, together with part (i) of Theorem 4.3, implies that $E(X) = m$. By (4.9),

$$E(U) = (2\pi)^{-1/2} \int_{-\infty}^{+\infty} ye^{-y^2/2} \, dy$$

which indeed equals zero because the integrand is symmetric about zero (and the integral is absolutely convergent).   ▲

Example 4.19 Let X and Y be independent normal variables with parameters $(m = 2, \ \sigma = 1)$ and $(m = -3, \ \sigma = 4)$, respectively. Let us evaluate $E(5XY - 8Y + 2)$.

From Example 4.18, $E(X) = 2$ and $E(Y) = -3$. Thus, by Theorem 4.3,

$$E(5XY - 8Y + 2) = E(5XY) + E(-8Y + 2) = 5E(X)E(Y) - 8E(Y) + 2$$
$$= -30 + 24 + 2 = -4 \qquad\qquad ▲$$

4.3 THE DISTRIBUTION AND THE EXPECTATION OF A FUNCTION OF RANDOM VARIABLE

Example 4.20 Let X be a discrete random variable with values $-2, 0, 1$ and 2, and with distribution $(0.1, 0.2, 0.4, 0.3)$. Let us evaluate $E(X^2)$.

Evidently, X^2 is a discrete random variable with values $4, 0$, and 1. Its distribution is $P(X^2 = 4) = P(X = -2) + P(X = 2) = 0.4$, $P(X^2 = 0) = P(X = 0) = 0.2$, and $P(X^2 = 1) = P(X = 1) = 0.4$. Hence

$$E(X^2) = 4 \times 0.4 + 0 \times 0.2 + 1 \times 0.4 = 2 \qquad\qquad ▲$$

Let us pause here for a moment. First note that $E(X) = -0.2 + 0.4 +$

$0.6 = 0.8$ and $E(X^2)$ are not related in any other way than that both are computed from the values and the distribution of X. That is, if only $E(X) = 0.8$ were known, we could not determine $E(X^2)$.

A second remark concerns the way in which $E(X^2)$ is computed. We had a choice there; we could have listed the value 4 twice for X^2, thus keeping the distribution of X unchanged; that is, we could have viewed X^2 whose values are 4, 0, 1, and 4 with distribution (0.1, 0.2, 0.4, 0.3). Of course, $E(X^2)$ would be the same, because the value 4 is now multiplied once with 0.1 and once with 0.3, so the aggregate contribution of 4 is 4×0.4, just as before. This second way of computing expectation is more convenient, because in this the readily available distribution of X can be utilized. This rule extends to other functions as well, not just to X^2, which is contained in the following statement.

A Convention on $g(X)$, When X Is Discrete: Let X be a discrete random variable with values x_1, x_2, ... and with distribution p_1, p_2, ... Let $g(x)$ be a function defined for all x_j. Then $Y = g(X)$ is a discrete random variable which can be interpreted in either of the following two ways: (i) The values of Y are $g(x_1)$, $g(x_2)$, ... (some values are possibly obtained several times), and the distribution of Y is the distribution p_1, p_2, ... of X. This interpretation immediately gives

$$(4.12) \qquad\qquad E(Y) = E(g(X)) = \sum g(x_j)p_j$$

whenever the sum above, taken over all j, is absolutely convergent. (ii) The second interpretation of Y is obtained when we insist on its values to be distinct. Hence Y is taking on the distinct values in the sequence $g(x_1)$, $g(x_2)$, ..., and its distribution is computed from the distribution of X by $P(Y = z) = \Sigma p_j$, where the summation is over those values of j for which $g(x_j) = z$. These two interpretations (or definitions) of Y are considered equivalent.

Example 4.21   Let us evaluate the expected value of (i) $X(X - 1)$ and (ii) e^{tX}, where t is a real number, when the distribution of X is (a) binomial, (b) Poisson, and (c) geometric.

(a) By (4.12), if X is binomial with parameters n and p,

$$E[X(X - 1)] = \sum_{m=0}^{n} m(m - 1)\binom{n}{m}p^m(1 - p)^{n-m} = n(n - 1)p^2$$

the detailed computation of which is given at (2.33). Furthermore,

$$E(e^{tX}) = \sum_{m=0}^{n} e^{tm}\binom{n}{m}p^m(1 - p)^{n-m} = \sum_{m=0}^{n} \binom{n}{m}(pe^t)^m(1 - p)^{n-m}$$

$$= (1 - p + pe^t)^n$$

by the binomial theorem [see, e.g., (2.22)].

(b) X is now Poisson with parameter λ. We have from (4.12) (note that the terms $k = 0$ and 1 below are 0)

$$E[X(X-1)] = \sum_{k=0}^{+\infty} k(k-1)\frac{\lambda^k}{k!}e^{-\lambda} = \lambda^2 e^{-\lambda}\sum_{k=2}^{+\infty}\frac{\lambda^{k-2}}{(k-2)!}$$

$$= \lambda^2 e^{-\lambda}e^{\lambda} = \lambda^2$$

where we utilized the Taylor formula for e^λ. Next,

$$E(e^{tX}) = \sum_{k=0}^{+\infty} e^{tk}\frac{\lambda^k}{k!}e^{-\lambda} = e^{-\lambda}\sum_{k=0}^{+\infty}\frac{(\lambda e^t)^k}{k!}$$

$$= e^{-\lambda}e^{\lambda e^t} = e^{\lambda(e^t-1)}$$

(c) We finally turn to the geometric distribution. In this case,

$$E[X(X-1)] = \sum_{k=1}^{+\infty} k(k-1)(1-p)^{k-1}p = p(1-p)\sum_{k=2}^{+\infty} k(k-1)(1-p)^{k-2}$$

In the last sum, we can recognize the second derivative of the sum [see (1.25)]

$$(4.13) \qquad\qquad \sum_{k=0}^{+\infty} x^k = \frac{1}{1-x} \qquad |x| < 1$$

at $x = 1 - p$. Thus, by differentiation,

$$E[X(X-1)] = p(1-p)\frac{d^{(2)}[1/(1-x)]}{dx^2}\Bigg|_{x=1-p} = \frac{2(1-p)}{p^2}$$

Applying (4.12) once more gives

$$E(e^{tX}) = \sum_{k=1}^{+\infty} e^{tk}(1-p)^{k-1}p = pe^t\sum_{k=1}^{+\infty}[e^t(1-p)]^{k-1}$$

$$= \frac{pe^t}{1-e^t(1-p)} \qquad t < -\log(1-p)$$

where the last equation is obtained from (4.13) with $x = e^t(1-p)$, which should be smaller than 1.   ▲

We now turn to the investigation of a function of an absolutely continuous random variable. Let X be a random variable with density function $f(x)$. Let $g(u)$ be a continuous function that is defined for all values of X. We investigate $Y = g(X)$. We start with an example, in which $g(u) = u^2$.

Example 4.22 Let X be a standard normal variable. Let us determine the density function and the expectation of X^2.

Evidently, for $z < 0$, $P(X^2 \le z) = 0$, and for $z > 0$,

$$P(X^2 \le z) = P(-\sqrt{z} \le X \le \sqrt{z}) = N(\sqrt{z}) - N(-\sqrt{z})$$

$$= 2N(\sqrt{z}) - 1 = 2(2\pi)^{-1/2} \int_{-\infty}^{z} e^{-t^2/2} \, dt - 1$$

where we applied $N(-x) = 1 - N(x)$. By differentiation, the density function of X^2, $g(z) = 0$ if $z < 0$, and

(4.14) $g(z) = 2(2\pi)^{-1/2} e^{-z/2} (\frac{1}{2} z^{-1/2}) = (2\pi z)^{-1/2} e^{-z/2}$ $z > 0$

We now compute the expectation by the formula of (4.9):

$$E(X^2) = \int_{-\infty}^{+\infty} xg(x) \, dx = (2\pi)^{-1/2} \int_{0}^{+\infty} x^{1/2} e^{-x/2} \, dx$$

Let us substitute $y^2 = x$. Then $2y \, dy = dx$, and thus

(4.15) $E(X^2) = 2(2\pi)^{-1/2} \int_{0}^{+\infty} y^2 e^{-y^2/2} \, dy = (2\pi)^{-1/2} \int_{-\infty}^{+\infty} y^2 e^{-y^2/2} \, dy$

from which, by integrating by parts [note that $y^2 e^{-y^2/2} = -y(e^{-y^2/2})'$],

$$E(X^2) = (2\pi)^{-1/2} \left\{ (-ye^{-y^2/2}) \right]_{-\infty}^{+\infty} + \int_{-\infty}^{+\infty} e^{-y^2/2} \, dy \right\} = 1 \qquad \blacktriangle$$

Note the following interpretation of (4.15). We transformed the random variable X into $Y = g(X)$ with $g(u) = u^2$. Yet the expectation of Y is finally computed by means of the density of X. This is achieved at the "expense" of the appearance of $g(y) = y^2$ in the integral instead of the customary y in an expectation. This is special neither to the function $g(y) = y^2$ nor to the normal density. The observation just described is formulated below as a general statement, but we do not go through its proof again. No change is required in the argument of Example 4.22 to prove the more general result.

Theorem 4.4. Let X be an absolutely continuous random variable with density function $f(x)$. Then for every continuous function $g(u)$, which is defined for all values of X,

$$E[g(X)] = \int_{-\infty}^{+\infty} g(x)f(x) \, dx$$

whenever the integral is absolutely convergent.

For calculating the distribution function or the density function of $g(X)$, the reader is advised to learn the technique of Example 4.22, rather than to memorize a formula. That is, in $P(g(X) \leq z)$, transform the inequality $g(X) \leq z$ to inequalities on X, and apply the distribution function of X to these new inequalities by which the distribution function $P(g(X) \leq z)$ of $g(X)$ is obtained. Differentiation of the distribution function gives the density function.

Example 4.23   The distribution function of X is the (standard) Pareto distribution function

$$F(z) = 1 - \frac{1}{z^c} \qquad z \geq 1, \quad c > 0$$

Let us determine the distribution function and the expectation of $Y = \log X$.

Since $P(X \leq 1) = F(1) = 0$, $X > 1$ with probability one. Consequently, $\log X > 0$, and thus $P(\log X \leq z) = 0$ for $z < 0$. On the other hand, for $z > 0$,

$$P(\log X \leq z) = P(X \leq e^z) = F(e^z) = 1 - e^{-cz} \qquad c > 0$$

That is, $\log X$ is exponentially distributed; we thus conclude from Example 4.13 that $E(\log X) = 1/c$.   ▲

Example 4.24   Let the distribution function of X be exponential with parameter $\lambda > 0$. Let us evaluate (i) $E(X^2)$ and (ii) $E(e^{tX})$, where t is a real number.

In both cases, we apply Theorem 4.4. By integrating by parts,

$$E(X^2) = \int_0^{+\infty} x^2 \lambda e^{-\lambda x}\, dx = -x^2 e^{-\lambda x}\Big]_0^{+\infty} + 2 \int_0^{+\infty} x e^{-\lambda x}\, dx$$

$$= \frac{2}{\lambda} \int_0^{+\infty} x \lambda e^{-\lambda x}\, dx = \frac{2}{\lambda^2}$$

where, in the last integral, Example 4.13 is applied. Next

$$E(e^{tX}) = \int_0^{+\infty} e^{tx} \lambda e^{-\lambda x}\, dx = \lambda \frac{e^{(t-\lambda)x}}{t - \lambda}\Big]_0^{+\infty} = \begin{cases} +\infty & \text{if } t \geq \lambda \\[2mm] \dfrac{\lambda}{\lambda - t} & \text{if } t < \lambda \end{cases} \qquad ▲$$

Example 4.25   Let X be a normal variable with parameters m and σ. We show that $E(X^2) = m^2 + \sigma^2$.

We know from (3.13) that there is a standard normal variable U such that $X = \sigma U + m$. Hence Theorem 4.3 and Examples 4.18 and 4.22 yield

$$E(X^2) = E(\sigma^2 U^2 + 2m\sigma U + m^2) = \sigma^2 E(U^2) + 2m\sigma E(U) + m^2 = \sigma^2 + m^2 \blacktriangle$$

Example 4.26   Let X be a normal variable with parameters m and σ. Let us show $E(e^{tX}) = \exp[tm + 1/2t^2\sigma^2]$.

We again utilize the representation (3.13), quoted in Example 4.25. In view of Theorems 4.3 and 4.4, we thus have

$$E(e^{tX}) = E(e^{t\sigma U + tm}) = e^{tm} E(e^{t\sigma U})$$

$$= e^{tm}(2\pi)^{-1/2} \int_{-\infty}^{+\infty} e^{t\sigma x} e^{-x^2/2} \, dx$$

$$= e^{tm+(1/2)t^2\sigma^2}(2\pi)^{-1/2} \int_{-\infty}^{+\infty} e^{-(x-t\sigma)^2/2} \, dx$$

which is the value claimed, since the last integral, together with $(2\pi)^{-1/2}$, is the integral of the normal density with parameters $t\sigma$ and 1. $\blacktriangle$

4.4   THE VARIANCE AND THE CORRELATION COEFFICIENT

Because the expected value E of a random variable X expresses the tendency of the arithmetical mean of copies of (observations on) X, we would like to measure the difference between X and E. The following concept is one of the several possibilities for doing this.

Definition 4.4   Let X be a random variable with finite expectation E. If the transformed random variable $(X - E)^2$ has finite expectation as well, then

$$V(X) = E[(X - E)^2]$$

is called the *variance* of X. The positive square root $\sigma(X) = V^{1/2}(X)$ is called the *standard deviation* of X.

Note that whether X is discrete or absolutely continuous, another form of $V(X)$ is as follows:

(4.16) $$V(X) = E(X^2) - E^2$$

That is, by Theorems 4.2 and 4.3,

$$V(X) = E(X^2 - 2EX + E^2) = E(X^2) - 2E[E(X)] + E^2$$

which is indeed (4.16).

The reader can now see that it was not an accident that we computed $E(X^2)$ for so many distributions in the examples of the preceding section. By collecting the results for $E(X^2)$ and for E for several distributions, we list their variance below [the computation is by (4.16)]. Note that, in some cases, we computed $E[X(X-1)]$, from which, by the relation $E[X(X-1)] = E(X^2 - X) = E(X^2) - E(X)$, $E(X^2)$ is easily computed.

(i) *The binomial distribution.* If X is binomial with parameters n and p, then $E(X) = np$ and $V(X) = np(1-p)$ (Examples 4.3 and 4.21).

(ii) *The Poisson distribution.* If X is Poisson with parameter $\lambda > 0$, then $E(X) = V(X) = \lambda$ (Examples 4.6 and 4.21).

(iii) *The geometric distribution.* If X is geometric with parameter p, then $E(X) = 1/p$ and $V(X) = (1-p)/p^2$ (Examples 4.4 and 4.21).

(iv) *The hypergeometric distribution.* If X is hypergeometric with parameters T, M, and t, then $E(X) = tM/T$ and

$$V(X) = t\frac{M}{T}\left(1 - \frac{M}{T}\right)\left(1 - \frac{t-1}{T-1}\right) \text{ (Example 4.11).}$$

(v) *The exponential distribution.* If X is exponentially distributed with parameter $\lambda > 0$, then $E(X) = 1/\lambda$ and $V(X) = 1/\lambda^2$ (Examples 4.13 and 4.24).

(vi) *The normal distribution.* If X is a normal variable with parameters m and σ, then $E(X) = m$ and $V(X) = \sigma^2$ (Examples 4.18 and 4.25).

Example 4.27   Let X be a random variable with finite expectation E and standard deviation σ. Let us compute $P(|X - E| < 2\sigma)$ if (i) X is exponentially distributed; and (ii) X is normal.

From the preceding list, if X is exponential with parameter λ, then $E = \sigma = 1/\lambda$. Hence

$$P(|X - E| < 2\sigma) = P\left(\frac{1}{\lambda} - \frac{2}{\lambda} < X < \frac{1}{\lambda} + \frac{2}{\lambda}\right) = P\left(X < \frac{3}{\lambda}\right)$$

$$= 1 - e^{-\lambda(3/\lambda)} = 1 - e^{-3} = 0.9502$$

where in the second step, we could drop $-1/\lambda < X$, because $X > 0$.

If X is normal with parameters m and σ, then $E = m$ and the standard deviation of X is in fact σ. Thus

$$P(|X - E| < 2\sigma) = P\left(-2 < \frac{X - m}{\sigma} < 2\right) \stackrel{.}{=} N(2) - N(-2)$$

$$= 2N(2) - 1 = 0.9544$$

where we utilized the fact that $(X - m)/\sigma$ is standard normal, and thus we could take $N(2)$ from Table 2.1.   ▲

Example 4.28   Let us compute the same probability as in Example 4.27, assuming that X is uniformly distributed over the interval $(0,1)$.

Since the uniform distribution is not included in the previous list of expectation and variance, we have first to compute E and σ. By definition, if X is uniform on $(0,1)$, then the density function of X,

$$f(x) = \begin{cases} 1 & \text{if } 0 < x < 1 \\ 0 & \text{otherwise} \end{cases}$$

and its distribution function $F(x) = x$ if $0 \le x < 1$, and $F(x) = 0$ or 1 according as $x < 0$ or $x \ge 1$. Therefore,

$$E = E(X) = \int_{-\infty}^{+\infty} xf(x)\,dx = \int_0^1 x\,dx = \frac{1}{2}$$

$$E(X^2) = \int_{-\infty}^{+\infty} x^2 f(x)\,dx = \int_0^1 x^2\,dx = \frac{1}{3}$$

and

$$V(X) = E(X^2) - E^2 = \frac{1}{3} - \frac{1}{4} = \frac{1}{12}, \quad \text{that is,} \quad \sigma = \left(\frac{1}{12}\right)^{1/2} = \frac{1}{2\sqrt{3}}$$

Thus the desired probability

$$P(|X - E| < 2\sigma) = P\left(\frac{1}{2} - \frac{1}{\sqrt{3}} < X < \frac{1}{2} + \frac{1}{\sqrt{3}}\right) = 1$$

because $1/\sqrt{3} > 1/2$, implying that $F(1/2 - 1/\sqrt{3}) = 0$ and $F(1/2 + 1/\sqrt{3}) = 1$.

▲

Because the variance is defined through a specific expectation, its basic properties are immediate from the basic properties of expectation.

Theorem 4.5 Let X be a random variable, discrete or absolutely continuous, whose expectation and variance are finite. Then
(i) $V(X) \ge 0$.
(ii) $V(X) = 0$ if, and only if, X is a constant.
(iii) $V(aX + b) = a^2 V(X)$ for all a and b.
    Furthermore, if Y is another random variable with finite expectation and variance, which is independent of X, then
(iv) $V(X + Y) = V(X) + V(Y)$.
Finally, if $X_1, X_2, \ldots, X_n$ are random variables, each with finite expectation and variance, and if, for every $i < j$, X_i and X_j are independent, then
(v) $V(X_1 + X_2 + \cdots + X_n) = V(X_1) + V(X_2) + \cdots + V(X_n)$.

Proof: (i) Because $V(X)$ is the expectation of $(X - E)^2 \ge 0$, it follows immediately from the definition of expectation, both in the discrete and the absolutely continuous case, that $V(X) \ge 0$.
    (ii) If $X = c$, a constant, we can view it as a discrete random variable

with the single value c and with distribution $P(X = c) = 1$. Hence $E(X) = c$ i.e., $X - E = 0$, and thus $V(X) = 0$ as well. Conversely, if X is not a constant, then $(X - E)^2$ is strictly positive with some probability, which contributes a positive amount to the expectation of $(X - E)^2$, i.e., to $V(X)$.

(iii) By Theorems 4.2 and 4.3, $E(aX + b) = aE(X) + b$, and thus $(aX + b) - E(aX + b) = a(X - E)$, from which

$$V(aX + b) = E([a(X - E)]^2) = a^2 E[(X - E)^2] = a^2 V(X)$$

(iv) Because $E(X + Y) = E(X) + E(Y)$, we can write

$$[X + Y - E(X + Y)]^2 = [X - E(X)]^2 + [Y - E(Y)]^2 + 2[X - E(X)][Y - E(Y)]$$

from which, by taking expectation (we can do it term by term), we get

$$V(X + Y) = V(X) + V(Y) + 2E[(X - E(X))(Y - E(Y))]$$

Now, since X and Y are independent, so are $X - E(X)$ and $Y - E(Y)$, and thus by Theorems 4.2 and 4.3,

$$E[(X - E(X))(Y - E(Y))] = E[X - E(X)]E[Y - E(Y)]$$

But $E[X - E(X)] = E(X) - E(X) = 0$ (so is, of course, the factor in terms of Y), concluding the claim that $V(X + Y) = V(X) + V(Y)$.

(v) This is just a repetition of the argument in (iv), whose details are therefore omitted.

The theorem is established.   ▲

Example 4.29   Let X be an exponential variable with $E(X) = 2$, and let Y be normal with parameters $m = -3$ and $\sigma = 2$. Let us determine $V(3X - 2Y + 5)$ if X and Y are independent.

Part (iii) of Theorem 4.5 implies that $V(3X - 2Y + 5) = V(3X - 2Y)$. Next, in view of the independence of X and Y, parts (iv) and (iii) of Theorem 4.5 yield

$$V(3X - 2Y) = V(3X) + V(-2Y) = 9V(X) + 4V(Y)$$

But we know that $V(X) = 4$ ($V(X) = E^2(X)$ for the exponential distribution), and $V(Y) = \sigma^2 = 4$, and thus $V(3X - 2Y + 5) = 9 \times 4 + 4 \times 4 = 52$.   ▲

In the course of the proof of part (iv) of Theorem 4.5, the only reference to independence was made when we established that

(4.17) $$E[(X - E(X))(Y - E(Y))] = 0$$

Although the converse is not true, that is, (4.17) may be valid even though X and Y are dependent, it became widespread in the literature to use the left-hand side of (4.17) as a measure of dependence. We thus introduce the following concept.

Definition 4.5 Let X and Y be random variables with finite expectation $E(X)$ and $E(Y)$, and standard deviation $\sigma(X)$ and $\sigma(Y)$, respectively. Then

$$\text{cov}(X,Y) = E[(X - E(X))(Y - E(Y))]$$

is called the covariance of X and Y, and the ratio

$$\text{corr}(X,Y) = \frac{\text{cov}(X,Y)}{\sigma(X)\sigma(Y)}$$

the correlation coefficient of X and Y. We call X and Y uncorrelated, if their covariance, and thus their correlation coefficient, is zero.

In the same way in which (4.12) and Theorem 4.4 were established, we get the following methods of computation of the covariance of X and Y.

First, let both X and Y be discrete with values $x_1, x_2, \ldots$ and $y_1, y_2, \ldots$, respectively, and let $p(i,j) = P(X = x_i, Y = y_j)$. Then, after $E(X)$ and $E(Y)$ have been computed,

$$(4.18) \qquad \text{cov}(X,Y) = \sum \sum (x_i - E(X))(y_j - E(Y))p(i,j)$$

where the summations are over all i and j.

Next, let both X and Y be absolutely continuous with joint density function $f(x,y)$. Then, again, after $E(X)$ and $E(Y)$ have been computed, we compute

$$(4.19) \quad \text{cov}(X,Y) = \int_{-\infty}^{+\infty} \int_{-\infty}^{+\infty} (x - E(X))(y - E(Y))f(x,y) \, dx \, dy$$

From (4.18) and (4.19), by the inequality known in calculus as the Cauchy-Schwarz inequality, we get

$$[\text{cov}(X,Y)]^2 \leq V(X)V(Y)$$

which can also be written as

$$(4.20) \qquad\qquad -1 \leq \text{corr}(X,Y) \leq 1$$

Example 4.30 Let X and Y be discrete random variables whose values and joint distribution are given in the following table:

X \ Y	−1	0	2
−1	0.1	0.2	0.05

| 0 | 0.05 | 0.1 | 0.1 |
| 2 | 0.2 | 0.1 | 0.1 |

Let us evaluate the covariance and the correlation coefficient of X and Y.

The first step is to determine the expectation of X and Y. For these, in turn, we have to know their distribution. With reference to Section 3.7 (recall, for example, Example 3.18), we utilize that row sums give the distribution of X, and column sums the distribution of Y. We thus have

$$P(X=-1)=0.35, \qquad P(X=0)=0.25, \qquad P(X=2)=0.4$$

$$P(Y=-1)=0.35, \qquad P(Y=0)=0.4, \qquad P(Y=2)=0.25$$

We can now easily compute $E(X)$, $V(X)$, $E(Y)$, and $V(Y)$:

$$E(X) = -1 \times 0.35 + 0 + 2 \times 0.4 = 0.45$$

$$V(X) = E(X^2) - E^2(X) = 0.35 + 0 + 4 \times 0.4 - (0.45)^2 = 1.7475$$

$$E(Y) = -1 \times 0.35 + 0 + 2 \times 0.25 = 0.15$$

$$V(Y) = E(Y^2) - E^2(Y) = 0.35 + 0 + 4 \times 0.25 - (0.15)^2 = 1.3275$$

From the variances, $\sigma(X) = 1.322$ and $\sigma(Y) = 1.152$, which we need in the computation of the correlation coefficient. We now apply (4.18) to compute $\text{cov}(X,Y)$. We first subtract 0.45 from the X values and 0.15 from the Y values, getting $(-1.45, -0.45, 1.55)$ for $x_i - E(X)$, $1 \le i \le 3$, and $(-1.15, -0.15, 1.85)$ for $y_j - E(Y)$, $1 \le j \le 3$. The product of these values, multiplied by the corresponding entry of the distribution table, will be the terms in $\text{cov}(X,Y)$, and the sum of all such terms is the desired covariance. The actual value is $\text{cov}(X,Y) = 0.0802$. This, divided by the earlier computed $\sigma(X)\sigma(Y)$, yields the correlation coefficient. That is,

$$\text{corr}(X,Y) = \frac{0.0802}{1.322 \times 1.152} = 0.053 \qquad \blacktriangle$$

Example 4.31   Let (X,Y) be a bivariate normal vector; that is, the joint density of X and Y is given by (3.28) and (3.29). Then, $\text{corr}(X,Y) = \rho$.

We first compute $\text{cov}(X,Y)$ by formula (4.19), in which $f(x,y)$ is given by (3.28) and (3.29). In fact, we put down this formula after substituting $u = (x - m_1)/\sigma_1$ and $v = (y - m_2)/\sigma_2$. With this substitution, the whole integral becomes free from m_1, m_2, σ_1, and σ_2, except that $\sigma_1\sigma_2$ comes to the front of the integral as a multiplier:

$$\text{cov}(X,Y) = C \int_{-\infty}^{+\infty} \int_{-\infty}^{+\infty} uv \, \exp\left[-\frac{1}{2(1-\rho^2)}(u^2 - 2\rho\, uv + v^2)\right] du \, dv$$

where

$$C = \frac{\sigma_1 \sigma_2}{2\pi\sqrt{1 - \rho^2}}$$

We write $u^2 - 2\rho uv + v^2 = (u - \rho v)^2 + (1 - \rho^2)v^2$. Hence the integral above becomes

$$\sigma_1 \sigma_2 (2\pi)^{-1/2} \int_{-\infty}^{+\infty} ve^{-v^2/2} \left[\int_{-\infty}^{+\infty} I(u,v) \, du \right] dv$$

with

$$I(u,v) = [2\pi(1 - \rho^2)]^{-1/2} \, u \, \exp\left[-\frac{1}{2(1 - \rho^2)}(u - \rho v)^2 \right]$$

We can recognize that $I(u,v)$ is u times the normal density with parameters ρv and $(1 - \rho^2)^{-1/2}$; consequently, the integral of $I(u,v)$ is the expected value of a normal variable with the just specified parameters. We found earlier (Example 4.18) that this expectation is ρv. That is,

$$\text{cov}(X,Y) = \sigma_1 \sigma_2 (2\pi)^{-1/2} \int_{-\infty}^{+\infty} \rho v^2 e^{-v^2/2} \, dv$$

This integral has come up before; since it can be written as $\sigma_1 \sigma_2 \rho E(U^2)$, where U is a standard normal variable, its value is $\sigma_1 \sigma_2 \rho$ (Example 4.22). We have thus obtained

$$\text{cov}(X,Y) = \sigma_1 \sigma_2 \rho$$

or, upon dividing by $\sigma_1 \sigma_2$, $\text{corr}(X,Y) = \rho$ ▲

Note that when $\rho = 0$ in the bivariate normal density, $f(x,y) = f_X(x)f_Y(y)$, where $f_X(x)$ and $f_Y(y)$ are the marginal (normal) densities; that is, the components X and Y are independent. Therefore, as a corollary to Example 4.31 we established that, for a bivariate normal vector (X,Y), the components X and Y are independent if, and only if, they are uncorrelated. Although the normal densities are not unique in this respect, only a few densities share this property.

4.5   THE CHEBYSHEV INEQUALITY AND THE (WEAK) LAW OF LARGE NUMBERS

The variance V was introduced to measure the deviation of a random variable X from its expectation E. It is therefore reasonable to ask whether we could estimate $P(|X - E| < u)$ by means of the variance V. Of course, if the distribution function $F(z)$ of X were known as well, the probability mentioned could be calculated (not just estimated) (see Examples 4.27 and 4.28). The answer is contained in the following theorem.

Theorem 4.6   (The Chebyshev Inequality)   If X is a random variable with finite expectation E and variance V, then for an arbitrary number $u > 0$,

$$(4.21) \qquad\qquad P(|X - E| \geq u) \leq \frac{V}{u^2}$$

or, equivalently,

$$(4.22) \qquad\qquad P(|X - E| < u) \geq 1 - \frac{V}{u^2}$$

The proof is based on the following inequality.

Lemma   (The Markov Inequality)   Let $Y \geq 0$ be a random variable with finite expectation $E_1 > 0$. Then, for an arbitrary number $t > 0$,

$$P(Y \geq tE_1) \leq \frac{1}{t}$$

Proof: First, let Y be discrete. If the values of Y are $y_1, y_2, \ldots$, and its distribution is $p_1, p_2, \ldots$, then, by assumption, each $y_j \geq 0$ and

$$E_1 = \sum y_j p_j < +\infty$$

where the summation is for all j. Consequently, if we sum $y_j p_j$ for some j only, we get a value smaller than E_1. Let us sum $y_j p_j$ over those values of j for which $y_j \geq tE_1$. Denoting this sum by $\sum^*$, we have

$$E_1 \geq \sum{}^* y_j p_j \geq \sum{}^* tE_1 p_j = tE_1 \sum{}^* p_j$$

Now, by the definition of $\sum^*$, it contains all those values of Y for which $Y \geq tE_1$; hence

$$\sum{}^* p_j = \sum{}^* P(Y = y_j) = P(Y \geq tE_1)$$

Substituting this into the preceding inequality, we get

$$E_1 \geq tE_1 P(Y \geq tE_1)$$

which, when divided by $tE_1 > 0$, is the inequality desired.

Next, we prove the lemma, when Y is absolutely continuous. Let its density function be $f(x)$. Since $Y \geq 0$, $f(x) = 0$ for $x < 0$, and thus

$$E_1 = \int_0^{+\infty} xf(x)\,dx < +\infty$$

The argument now is the same as in the discrete case, except that instead of sums, we deal with an integral. That is, since $tE_1 > 0$,

$$E_1 \geq \int_{tE_1}^{+\infty} xf(x)\, dx \geq \int_{tE_1}^{+\infty} tE_1 f(x)\, dx = tE_1 \int_{tE_1}^{+\infty} f(x)\, dx$$

$$= tE_1[1 - F(tE_1)] = tE_1 P(Y \geq tE_1)$$

where $F(z)$ is the distribution function of Y. The extreme sides, after dividing by tE_1, give the desired inequality. The proof is complete. ▲

Proof of Theorem 4.6 Put $Y = (X - E)^2$. Then $Y \geq 0$, and $E_1 = E(Y) = V(X)$. By Theorem 4.5, $V(X) \geq 0$, and, in fact, $V(X) > 0$, except when X is a constant c. However, if $X = c$, then $E = c$ as well, and thus both sides of (4.21) are zero. Therefore, in the following discussion, we can assume that X is not a constant, implying that $V(X) > 0$. Hence the lemma is applicable, yielding

$$P[(X - E)^2 \geq tV] \leq \frac{1}{t}$$

or, equivalently,

(4.23) $$P(|X - E| \geq (tV)^{1/2}) \leq \frac{1}{t}$$

Let $u = (tV)^{1/2}$. Then $t = u^2/V$; that is, (4.23) is just another form of (4.21). If we turn to the complement of the event in (4.21), we get (4.22), which completes the proof. ▲

Recall Section 2.6, where a Chebyshev inequality is presented for X with binomial distribution having parameters n and p. Since, in that case, $E(X) = np$ and $V = np(1 - p)$, the inequality (2.24) is, in fact, (4.21) for the binomial distribution.

Example 4.32 Let X be a random variable with expectation $E = 2$ and variance $V = 4$. Let us estimate $P(|X - 2| < 4)$.
From (4.22) with $u = 4$,

$$P(|X - 2| < 4) \geq 1 - \frac{4}{16} = 0.75 \qquad ▲$$

In addition to its usefulness in computations, the Chebyshev inequality has the following very significant theoretical consequence.

Theorem 4.7 (The Weak Law of Large Numbers) Let $X_1, X_2, \ldots, X_n$ be independent random variables with the same finite expectation E and variance V. Then, for every $\varepsilon > 0$, as $n \to +\infty$,

$$\lim P\left(\left|\frac{X_1 + X_2 + \cdots + X_n}{n} - E\right| \geq \varepsilon\right) = 0$$

Remarks (1) Note that V does not appear in the conclusion. In fact, it would not be needed among the assumptions either. However, since we prove Theorem 4.7 as a consequence of the Chebyshev inequality, we had to include the finiteness of V.

(2) The most important application of the weak law of large numbers is the case when the X_j are assumed to be independent copies of a random variable whose expectation is E (and whose variance is V). We then have that for sufficiently large n, the arithmetical mean of the X_j is as close to E as we wish ($\varepsilon > 0$ is chosen by us in advance) with probability close to 1. This was our argument at the introduction of the concept of expectation, except that we could not specify accurately what we meant by "being asymptotically equal" (recall Theorem 4.1). This is now made accurate in Theorem 4.7.

Proof: Put $M_n = (1/n)(X_1 + X_2 + \cdots + X_n)$. Then, by Theorems 4.2 and 4.3,

$$E(M_n) = \frac{1}{n} E(X_1 + X_2 + \cdots + X_n) = \frac{1}{n} \left[E(X_1) + E(X_2) + \cdots + E(X_n)\right] = E$$

and a similar computation from Theorem 4.5 leads to $V(M_n) = (1/n)V$. Hence, by the Chebyshev inequality, for an arbitrary $\varepsilon > 0$,

(4.24) $$P(|M_n - E| \geq \varepsilon) \leq \frac{V}{n\varepsilon^2}$$

which indeed goes to zero as $n \to +\infty$.   ▲

Note that (4.24) is stronger than the statement of the theorem in that it can be utilized in computations for given n, E, and V. The reader will find questions in this regard among the exercises, which do not require more than a straight substitution into (4.24).

4.6   EXERCISES

1. Find the expected value of the discrete random variable X with $P(X = -2) = 0.2$, $P(X = 0) = 0.3$, $P(X = 1) = 0.4$, and $P(X = 5) = 0.1$.

2. An urn contains three white, two red, and two blue balls. Assume that five balls are selected from the urn at random without replacement. Find the expected number of (i) white, (ii) red, and (iii) blue balls among the five balls selected.

3. Roll a die repeatedly, and stop either at the fifth roll, or when a 6 comes up, whichever occurs first. Find the expected number of rolls.

4. At a roulette table with 38 numbers, one wins 36 to 1 if the winning number is picked. Find the expected profit of the casino on 1 million bets of $1 each on single numbers.

5. Assume that the probability that someone agrees with your views is 1/3. If you ask people at random (independently of each other), what is the expected number of people you should ask until you meet someone who agrees with you? How does this change if 1/3 is changed to 0.4? What is the meaning of either of these expectations?

6. A number is picked at random from each of the following sets: $\{1,2,3\}$, $\{1,3,4\}$, $\{2,3,5,6\}$, and $\{1,2\}$. Find the expected number of odd integers selected.

7. The random variable X is Poisson with parameter $\lambda = 2$. Find $E(3 - 5X)$.

8. In a Poisson process, the expected number of points in any interval of unit length is 3. Let X be the number of points in the interval $(3, 5.5)$. Find $E(X)$ and $P(X = 2)$.

9. If X is binomial variable with parameters $n = 10$ and $p = 0.3$, and Y is a Poisson variable with $\lambda = 2$, find $E(3X - 5Y + XY)$ if X and Y are independent.

10. From a well-shuffled deck of cards, the cards are turned face up, one by one. What is the expected number of cards that have to be turned up in order to obtain two kings?

11. A shelf has 10 items, of which 3 are defective. A customer purchases four of these items. What is the expected number of defective items that the person has purchased?

12. The density function of X,

$$f(x) = \begin{cases} 3x^2 & \text{if } 0 < x < 1 \\ 0 & \text{otherwise} \end{cases}$$

Find $E(X)$ and $E(5 - 3X)$.

13. Assume that customers arrive at a store according to a Poisson process. If the expected time to the arrival of the first customer is 5 minutes, what is the expected number of customers in an hour?

14. The random variable X is normally distributed with $E(X) = -1$ and $\sigma = 3$. Use Table 2.1 to compute $P(X \le 4)$.

15. The density function of the vector (X, Y),

$$f(x,y) = \begin{cases} Cxy & \text{if } 0 < x < y < 2 \\ 0 & \text{otherwise} \end{cases}$$

Determine (i) C; (ii) the marginal densities of X and Y; and (iii) $E(X)$ and $E(Y)$.

16. The random variables X and Y are independent, where X is exponential with parameter $\lambda = 1/2$ and Y is normal with parameters $m = \sigma = 2$. Find $E(3X - 5Y + 2XY)$.

17. Find $E(X^2)$, $E(X^3)$, and $E(e^{tX})$ for X of Exercise 1.

18. Find $E(X^2)$, $E(X^3)$, and $E(e^{tX})$ for X of each of Exercises 12 and 15.

19. Let X be a normal variable with $E(X) = 2$ and $E(X^2) = 8$. Use Table 2.1 to compute $P(-1 < X < 4)$.

20. Evaluate the variance of X of Exercise 1.

21. Find the variance of the white and red balls obtained in Exercise 2.

22. Find the variance of X of Exercise 12.

23. Find the variance of X of Exercise 15.

24. Is it possible that the waiting time to the first arrival in a Poisson process has expected value 3 and variance 5?

25. Assume that the weight X in kilograms of male students is a normal random variable with expectation 76 and variance 36. Find $P(68 < X < 86)$.

26. If X is normal with parameters $m = -2$ and $\sigma = 2$, what is the variance of $3 - 5X$?

27. Find the variance of $5X - 3Y - 12$, where X and Y are independent, X is exponentially distributed with $E(X) = 2$, and Y is normal with parameters $m = 3$ and $\sigma = 0.5$.

28. Find $\text{cov}(X,Y)$ for the vector (X,Y) of Exercise 15.

29. Let X be the number of 1s and Y the number of 6s in 12 rolls of a fair die. Find $\text{cov}(X,Y)$.

30. Find $\text{cov}(X, X + Y)$, where X and Y are independent exponential variables with parameters $\lambda_1 = 1/2$ and $\lambda_2 = 2$, respectively.

31. Find $\text{corr}(X,Y)$ for (X,Y) of each of Exercises 28 and 29.

32. Write the density function of (X,Y) if it is bivariate normal with $E(X) = -2$, $E(Y) = 3$, $V(X) = 1$, $V(Y) = 4$, and $\text{cov}(X,Y) = 1.5$.

33. Estimate by the Markov inequality $P(X \geq 4)$, if X is exponentially distributed with parameter $\lambda = 2$.

34. Estimate $P(24 < X < 40)$, if X is a random variable with $E(X) = 32$ and $V(X) = 4$.

35. The (random) profit on a product item is 2 (in some units) if it is of first quality, 0.5 if it should be sold at a discount due to some imperfection, and -0.8 if it is defective. The company produces 1600 of these items. Past experience shows that an item is of first quality, discounted, and defective, respectively, with probabilities 0.4, 0.55, and 0.05. Use the Chebyshev inequality to advice the company concerning its profit picture.

36. Let the distribution functions of both X and Y be either discrete or absolutely continuous. Assume both $E(X)$ and $E(Y)$ are finite. Show that $X \leq Y$ (with probability 1) implies $E(X) \leq E(Y)$.

5
Limit Theorems

5.1 THE CENTRAL LIMIT THEOREM

For a random variable X with finite expectation E and variance V, we established in Chapter 4 that if $X_1, X_2, \ldots, X_n$ are independent copies of X, then

$$(5.1) \qquad P\left(\left|\frac{X_1 + X_2 + \cdots + X_n}{n} - E\right| < \varepsilon\right) \geq 1 - \frac{V}{n\varepsilon^2}$$

where $\varepsilon > 0$ is an arbitrary number. Now, as $n \to +\infty$, the right-hand side, and thus the left-hand side as well, approaches 1, yielding a limit theorem. This limit theorem, however, is weak in the sense that it is deduced from an inequality, and thus the convergence of the right-hand side to 1 might provide a crude picture of the left-hand side as to its asymptotic behavior. The following theorem corrects this by giving an exact formula for the asymptotic behavior of the left-hand side of (5.1).

Theorem 5.1 (The Central Limit Theorem) With the notations of the preceding paragraph, for arbitrary $a < b$ (possibly $a = -\infty$ or $b = +\infty$), as $n \to +\infty$,

$$(5.2) \quad \lim P\left(a\sqrt{\frac{V}{n}} < \frac{X_1 + X_2 + \cdots + X_n}{n} - E < b\sqrt{\frac{V}{n}}\right) = N(b) - N(a)$$

where $N(z)$ is the standard normal distribution function, and thus its values can be taken from Table 2.1.

Remarks   (1) For comparing (5.1) and (5.2), let us choose $a = -b$ with $b > 0$ finite, and let $\varepsilon = b(V/n)^{1/2}$. Then the probabilities on the left-hand sides of (5.1) and (5.2) are identical, for which (5.1) gives the lower estimate $1 - 1/b^2$, and (5.2) gives the asymptotic value $N(b) - N(-b) = 2N(b) - 1$ [recall that $N(-b) = 1 - N(b)$]. By looking at several entries of Table 2.1, we can immediately see that $2N(b) - 1$ is considerably larger than $1 - 1/b^2$, implying that (5.1) is indeed crude as a limit theorem. However, it should not be forgotten that (5.1) is applicable with an arbitrary n, and thus when n is not large enough to apply (5.2), then (5.1) is the only formula available.

(2) This brings up the usual question: How large should n be for (5.2) to be applicable? The answer to this question is not straightforward, because it depends on the distribution function of X, which does not take place in the theorem itself. For most distributions, however, a rule of thumb is that (5.2) is applicable for $n \geq 50$, if nV is "not too small" (e.g., if $nV \geq 5$, which is, by no means, the best general rule).

(3) The normal approximation to the binomial (Section 2.6) is a special case of Theorem 5.1, because a binomial variable can be represented as the sum of independent indicator variables (Example 4.10).

(4) Sometimes it is more convenient to do the computations if we bring (5.2) into the form

$$(5.3) \qquad \lim P\left(a < \frac{X_1 + X_2 + \cdots + X_n - nE}{\sqrt{nV}} < b\right) = N(b) - N(a)$$

(5) There is a very interesting nonmathematical consequence of Theorem 5.1. In nonmathematical terms it says that when we observe a random phenomenon (which has finite expectation and variance) in a large quantity, we always end up with the normal distribution. It therefore suggests (and empirical studies confirm it) that "most distributions" in nature are normal (e.g., as the number of insects are normally distributed in an area, as are stars in the sky, the trees in a forest, etc.). The belief that there is only one observable random law in nature was broken, however, when it was discovered that there are several versions of the central limit theorem (in which, of course, the variables X_j are either not identically distributed, or they do not have finite variance or even finite expectation).

The proof of Theorem 5.1 is given in the next section. The rest of the present section is devoted to applications of the central limit theorem.

Example 5.1 Let $X_1, X_2, \ldots, X_{100}$ be independent and identically distributed random variables with expectation $E = 2$ and variance $V = 1/4$. Let us apply the central limit theorem to approximate

$$P(192 < X_1 + X_2 + \cdots + X_{100} < 210)$$

We apply (5.3). Since $nE = 200$ and $nV = 25$,

$$P(192 < X_1 + \cdots + X_{100} < 210)$$

$$= P\left(\frac{192 - 200}{5} < \frac{X_1 + \cdots + X_{100} - 100E}{\sqrt{100V}} < \frac{210 - 200}{5}\right)$$

that is, $b = 2$ and $a = -1.6$, and thus the approximation is

$$N(2) - N(-1.6) = N(2) - [1 - N(1.6)] = 0.9772 - 0.0548 = 0.9224$$

where the values $N(2)$ and $N(1.6)$ were taken from Table 2.1. ▲

Example 5.2 Assume that customers arrive at a bank according to a Poisson process with expectation 1.6 per minute. Denoting by $X(t)$ the number of customers arriving in a time interval of length t (measured in minutes), let us approximate $P(175 < X(120) < 215)$ by the central limit theorem.

Recall that, in a Poisson process, arrivals in nonoverlapping intervals are assumed to be independent random variables, and arrivals during time intervals of the same length are identically distributed (Section 3.11). Hence the number $X(120)$ of arrivals in a 2-hour period is the sum of 120 independent copies of $X(1)$. Since the distribution of $X(t)$ is Poisson with parameter λt(Theorem 3.10), and we know that $E(X(1)) = 1.6$, we have $\lambda = 1.6$. That is, $X(120) = X_1(1) + X_2(1) + \cdots + X_{120}(1)$, $E(X(120)) = 120 \times 1.6 = 192$ and $V(X(120)) = 120 \times 1.6 = 192$ (for a Poisson variable $E = V = \lambda$). Thus, by (5.3) and Table 2.1,

$$P(175 < X(120) < 215) = P\left(\frac{175 - 192}{\sqrt{192}} < \frac{X(120) - 120E}{\sqrt{120V}} < \frac{215 - 192}{\sqrt{192}}\right)$$

$$\sim N(1.66) - N(-1.23) = N(1.66) - [1 - N(1.23)]$$

$$= 0.9515 - 0.1093 = 0.8422 \qquad ▲$$

Example 5.3 Assume that the time length of servicing a customer at a bank is exponentially distributed with expected value 3 (minutes). Let Y be the aggregate time spent by the tellers on servicing 200 customers. Let us evaluate $P(Y > 660)$ through the normal approximation contained in the central limit theorem.

Let X_j be the time length needed to service the jth customer. Then the X_j can be assumed to be independent, and each X_j is an exponential variable with parameter $\lambda = 1/3$ $[E(X_j) = 1/\lambda = 3]$. Now, $Y = X_1 + X_2 + \cdots + X_{200}$, and thus, by (5.3),

$$P(Y > 660) = P\left(\frac{Y - 200E}{\sqrt{200V}} > \frac{660 - 600}{\sqrt{1800}}\right) = 1 - N(\sqrt{2}) = 1 - N(1.41)$$
$$= 0.0793$$

where Table 2.1 was again applied.   ▲

Example 5.4   A typical homeowner's insurance policy provides coverage of loss incurred due to accident on the premises of the covered home. We know from Section 3.3 that the time length X up to the occurrence of an accident is exponentially distributed. Therefore, the insurance company, after determining the single parameter λ of X, can compute the probability of an accident in a home in a year, which can then be used to predict the frequency of accidents among all policyholders. However, how can λ be determined?

Because $E(X) = 1/\lambda$ for the exponential distribution, the weak law of large numbers suggests that the arithmetical mean of a large number of observations can be taken as an approximation to $1/\lambda$. This seems to be a perfect mathematical solution, which is further enhanced by the fact that the error in this approximation can be calculated with the help of the central limit theorem. However, since a typical X is around 30 years, the whole approach is very impractical. Luckily, there are ways of getting around such difficulties, one of which has, in fact, been discussed before. Namely, recall our examples involving electric light bulbs, in one of which we have shown (Example 4.15) that in a group of k independent exponential variables, the smallest one is also exponential, whose parameter is k times the original parameter. Therefore, the insurance company can collect its records on X by grouping, say, 50 policyholders together, and use only the first accident in the group for estimating the unknown parameter λ (the arithmetical mean of these observations will be $1/50\lambda$). This approach is now practical because if $1/\lambda$ is about 30 years, then $1/50\lambda$ is about $30/50 = 0.6$ year; that is, accidents can be expected to occur as often as 0.6 year per group.   ▲

It was intentional not to put calculations into this example. Its aim was to show that in the course of applying probability theory, the nicest and most powerful mathematical results can become impractical. At the same time, we have shown that by searching for additional results, we could settle the problem of impracticality.

Example 5.5   Assume that a constant c in an equation in physics can be determined only through experiments. Because the experiments are not exact, not c, but a random quantity X, is measured whose expectation $E(X) = c$. If the randomness of X is due not only to inaccuracies of technique and instrument, but also to outside physical effects which cannot be eliminated from the experiment, the distribution of X cannot be determined. Assume that a physicist knows from prior estimates that the variance of X in the experiments is 0.25, and would like to determine c through the arithmetical mean of n observations $X_1, X_2, \ldots, X_n$ with the following accuracy:

$$P\left(\left|\frac{X_1 + X_2 + \cdots + X_n}{n} - c\right| < 0.1\right) \geq 0.975$$

How many observations should this physicist make?

We apply (5.2) with $a = -b$. Then $N(b) - N(a) = 2N(b) - 1$, and thus, by the accuracy chosen, b should be such that $2N(b) - 1 \geq 0.975$. Table 2.1 yields $b \geq 2.24$. Therefore, $2.24(0.25/n)^{1/2} = 0.1$, from which $n \geq 11.2^2 = 125.44$. That is, $n \geq 126$ will guarantee the accuracy required for c.   ▲

Example 5.6 Let Y be a Poisson variable with a positive integer parameter λ. Let us deduce from the central limit theorem that as $\lambda \to +\infty$,

(5.4) $$\lim P(Y \leq \lambda + \sqrt{\lambda}z) = N(z)$$

Because sums of independent Poisson variables are Poisson (Example 3.33), Y can be represented as $Y = X_1 + X_2 + \cdots + X_\lambda$, where the X_j are independent Poisson variables, each with parameter 1. Since $E(X_j) = V(X_j) = 1$, (5.3) with $a = -\infty$ becomes (5.4).   ▲

The limit relation (5.4) remains valid even if $\lambda > 0$ is not an integer, which can be proved either directly (by the method of the next section), or can be deduced from Example 5.6. Namely, if $\lambda = n + t$, where n is an integer and $0 < t < 1$, Y can still be written as $Y = X_1 + X_2 + \cdots + X_n + X_t^*$, where the X_j and X_t^* are independent Poisson variables, and each X_j has parameter 1, but X_t^* has parameter t. Now, if we write

$$(Y < \lambda + \sqrt{\lambda}z) = \left(\frac{Y - \lambda}{\sqrt{\lambda}} < z\right) = \left(\frac{X_1 + X_2 + \cdots + X_n - n}{\sqrt{\lambda}} + \frac{X_t^* - t}{\sqrt{\lambda}} < z\right)$$

it is not difficult to show that, since $(X_t^* - t)/\sqrt{\lambda} \to 0$ as $\lambda \to +\infty$, this term has no influence on the asymptotic behavior of $(Y - \lambda)/\sqrt{\lambda}$. Hence only the first term matters on the extreme right-hand side above, which, by the central limit theorem, becomes asymptotically normal.

The relation (5.4) is very useful, because it assures that a table of limited size suffices for the distribution function of a Poisson variable. Usually, with $\lambda \geq 25$, one can compute the Poisson distribution function from a normal table. The normalization by λ and $\sqrt{\lambda}$, however, must not be forgotten.

Example 5.7   Let Y be a Poisson variable with parameter $\lambda = 36$. Let us evaluate $P(Y < 45)$.

We write $45 = \lambda + \sqrt{\lambda}z = 36 + 6z$, from which $z = 1.5$. Hence, by (5.4) and Table 2.1, $P(Y < 45) \sim N(1.5) = 0.9332$.   ▲

The asymptotic normality of a distribution function, which is known to be the distribution function of a sum of independent and identically distributed random variables, can be established in the same way in which Example 5.6 was deduced from the central limit theorem. Among these we mention the following two cases, distributions that play a prominent role in statistics.

The Gamma Distribution: A random variable X has a *gamma distribution* (function), or we call X a *gamma variable*, if its density function $f(z) = 0$ for $z < 0$, and

(5.5) $$f(z) = C\lambda^a z^{a-1} e^{-\lambda z} \qquad z \geq 0 \quad (a > 0, \lambda > 0)$$

where C is a constant, depending on a alone and such that the integral of $f(z)$ equals 1. (The reciprocal of this constant is also known as the gamma function at a.) Note that the case $a = 1$ reduces to the exponential density function.

Now, for two independent gamma variables X_1 and X_2 with the same λ parameter but a possibly different a parameter, $X_1 + X_2$ also has a gamma distribution. This claim, of course, then extends to $X_1 + X_2 + \cdots + X_n$ for an arbitrary n. In particular, the sum of independent exponential variables with the same parameter is a gamma variable. All these can easily be obtained from Theorem 3.9. From the central limit theorem we can thus deduce that if X is a gamma variable, and if $a \to +\infty$, then $(X - A)/B$, with appropriate numbers A and B, is asymptotically standard normal.

The Chi-square Distribution: Let $X_1, X_2, \ldots, X_n$ be independent standard normal variables, and put $Y = X_1^2 + X_2^2 + \cdots + X_n^2$. The distribution (function) of Y is called a *chi-square distribution* (function) with n degrees of freedom.

It turns out that a chi-square distribution is a special gamma distribution. Thus, by comparing the density function (5.5) with the density function obtained for X_1^2 at (4.14), we find that they coincide with $a = 1/2$ and $\lambda = 1/2$. Consequently, by the property concerning the sums of gamma variables it follows that a chi-square density function is a gamma density function whose parameters are of the form $a = 1/2n$ with a positive integer n, and $\lambda = 1/2$. Their separation, both in name and treatment, is primarily for historical reasons.

5.2 MOMENT GENERATING FUNCTIONS, AND THE PROOF OF THE CENTRAL LIMIT THEOREM

The moment generating function of a random variable X is defined by

$$(5.6) \qquad M(t) = M_X(t) = E(e^{tX})$$

for all real numbers t such that the expectation above is finite. In other words, if X is discrete with values $x_1, x_2, \ldots$ and with distribution $p_1, p_2, \ldots$, then

$$(5.7) \qquad M(t) = \sum p_j \exp(tx_j)$$

whenever it is finite, where summation is over all values of j. On the other hand, if X is absolutely continuous with density function $f(z)$, then

$$(5.8) \qquad M(t) = \int_{-\infty}^{+\infty} e^{tz} f(z)\, dz$$

whenever the integral is finite.

The reason $M(t)$ is called the moment generating function of X is that if we evaluate the successive derivatives of $M(t)$ and set $t = 0$, we get (interchanging summation and differentiation, or integration and differentiation, as it may apply, is justified whenever the final form is absolutely convergent):

(i) In the discrete case

$$M'(0) = \sum x_j p_j; \quad M''(0) = \sum x_j^2 p_j; \quad \ldots; \quad M^{(k)}(0) = \sum x_j^k p_j$$

(ii) In the absolutely continuous case

$$M'(0) = \int_{-\infty}^{+\infty} z f(z)\, dz; \quad M''(0) = \int_{-\infty}^{+\infty} z^2 f(z)\, dz; \quad \ldots;$$

$$M^{(k)}(0) = \int_{-\infty}^{+\infty} z^k f(z)\, dz$$

which can be combined as

$$(5.9) \qquad M^{(k)}(0) = E(X^k) \qquad k = 1, 2, \ldots$$

These expected values have the common name of *moments* of X (thus the first moment is the expectation). So if the moments of X are finite, they can be computed from $M(t)$ by differentiation. Conversely, if the moments of X are finite, $M(t)$ can be expanded into a Taylor expansion

$$(5.10)\, M(t) = 1 + M'(0)t + M''(0)\frac{t^2}{2} + \cdots = 1 + E(X)t + E(X^2)\frac{t^2}{2} + \cdots$$

where as many terms can be used as many moments are finite.

Among the examples of Chapter 6, we have computed the moment generating function for several distributions (without referring to them as such). For easier reference, we restate some of those results:

Binomial distribution: $M(t) = (1 - p + pe^t)^n$ (all t)
Poisson distribution: $M(t) = \exp[\lambda(e^t - 1)]$ (all t)

Exponential distribution: $M(t) = \dfrac{\lambda}{\lambda - t}$ $(t < \lambda)$

Standard normal distribution: $M(t) = e^{t^2/2}$ (all t)

Example 5.8   Using the moment generating function, let us determine the moments of an exponential variable X with parameter λ.
   We apply (5.9). Since $M(t) = \lambda(\lambda - t)^{-1}$, successive differentiation yields

$$M^{(k)}(t) = \lambda(k!)(\lambda - t)^{-k-1}$$

With the substitution $t = 0$ we thus get

$$E(X^k) = M^{(k)}(0) = k!\lambda^{-k}$$

Earlier we found these values for $k = 1$ and 2.   ▲

Because the moment generating function is defined through the concept of expectation, the following properties are immediate from the results of Chapter 4.

Theorem 5.2   (i) If a and b are constants, then

$$M_{aX+b}(t) = E(e^{(aX+b)t}) = E(e^{aXt}e^{bt}) = e^{bt}M_X(at)$$

(ii) If $X_1, X_2, ..., X_n$ are independent random variables, then

$$M_{X_1+X_2+\cdots+X_n}(t) = E\{\exp[(X_1 + X_2 + \cdots + X_n)t]\}$$

$$= E(\prod_{j=1}^{n} e^{X_j t}) = \prod_{j=1}^{n} M_{X_j}(t)$$

Another important property of the moment generating function is stated, without proof, in the next theorem.

Theorem 5.3   Let Y_j, $j \geq 1$, be a sequence of random variables with distribution functions $F_j(z)$ and moment generating functions $M_j(t)$, $j \geq 1$. Assume that each $M_j(t)$ is defined for all t, and $M_n(t) \to M(t)$ for all t, as

$n \to +\infty$, where $M(t)$ is the moment generating function of the distribution function $F(z)$. Then, as $n \to +\infty$, $F_n(z) \to F(z)$ for all z where $F(z)$ is continuous.

Theorems 5.2 and 5.3 provide the necessary tools for proving the central limit theorem.

Proof of Theorem 5.1:   We prove Theorem 5.1 under the additional assumption that the common moment generating function $M(t)$ of the X_j is defined for all t. Let us add here that this extra assumption could be avoided if we were willing to deal with functions of a complex variable, namely, if in (5.6) – (5.8) we would replace t by a complex number. So, not the theorem, but our way of its proof requires the additional assumption.
   Put

$$X_{av} = \frac{1}{n}(X_1 + X_2 + \cdots + X_n)$$

It is sufficient to prove that

(5.11) $$F_n(z) = P\left(X_{av} - E \leq z\sqrt{\frac{V}{n}}\right) \to N(z)$$

because it then implies that $F_n(b) - F_n(a) \to N(b) - N(a)$ for all $a < b$; that is, (5.2) holds (note that it does not matter whether we permit equality sign or not on the left hand side of (5.2) because $N(z)$ is continuous at every z). Now, applying Theorem 5.3 with $Y_n = (X_{av} - E)(n/V)^{1/2}$, (5.11) will be established if we show that the moment generating function $M_n(t)$ of Y_n converges to $\exp(t^2/2)$ [which is the moment generating function of $N(z)$].
   Let us evaluate $M_n(t)$ by an appeal to Theorem 5.2. We get

$$M_n(t) = M_{X_{av}-E}\left(\sqrt{\frac{n}{V}}t\right) = \exp\left(-Et\sqrt{\frac{n}{V}}\right)M_{X_{av}}\left(\sqrt{\frac{n}{V}}t\right)$$

$$= \exp\left(-Et\sqrt{\frac{n}{V}}\right)M_{X_1+X_2+\cdots+X_n}\left(\sqrt{\frac{n}{V}}\frac{t}{n}\right) = e^{-Et(n/V)^{1/2}}M^n\left(\frac{t}{\sqrt{nV}}\right)$$

By applying (5.10) with the t^2 term fully used, the error term will be of smaller order of magnitude than the last major term, t^2 in this case, which is usually expressed by the symbol $o(t^2)$. That is,

$$M(t) = 1 + Et + (V + E^2)\frac{t^2}{2} + o(t^2)$$

where we utilized $E(X^2) = V(X) + E^2(X)$. Note that in the last expression for $M_n(t)$ above, $M(t/\sqrt{nV})$ appears, which when replaced by the quoted Taylor expansion, yields

$$M_n(t) = e^{-Et(n/V)^{1/2}} \left[1 + E\frac{t}{\sqrt{nV}} + (V + E)^2 \frac{t^2}{2nV} + o\left(\frac{t^2}{n}\right) \right]^n$$

If we take the logarithm and utilize the fact that $\log(1 + z) = z - (1/2)z^2 + o(z^3)$ as $z \to 0$, a somewhat lengthy but routine calculation yields

$$\log M_n(t) = \frac{t^2}{2} + o(1)$$

where $o(1)$ means "smaller magnitude than 1" (i.e., it converges to zero as $n \to +\infty$). In other words,

$$M_n(t) \to e^{t^2/2}$$

which completes the proof.   ▲

It is left as an exercise to give a new proof for the Poisson approximation to the binomial distribution by means of moment generating functions.

5.3   ASYMPTOTIC EXTREME VALUE DISTRIBUTIONS

Let $X_1, X_2, \ldots, X_n$ be independent and identically distributed random variables with (common) distribution function $F(z)$. Put

$$W_n = \min(X_1, X_2, \ldots, X_n), \qquad Z_n = \max(X_1 X_2, \ldots, X_n)$$

and

$$L_n(z) = P(W_n \le z), \qquad H_n(z) = P(Z_n \le z)$$

Because $\{W_n > z\}$ means that each $X_j > z$, and $\{Z_n \le z\}$ means that each $X_j \le z$, the assumption of independence implies that

$$L_n(z) = 1 - P(W_n > z) = 1 - [1 - F(z)]^n$$

and

$$H_n(z) = F^n(z)$$

It has been analyzed in Section 3.11 that even very good approximations to $F(z)$ give distorted values for $H_n(z)$ (because F^n is very sensitive to changes in F). Therefore, when an exact $F(z)$ is not available, a good approximation to $H_n(z)$ should be found not in terms of an approximation to $F(z)$, but by means of the whole $F^n(z)$, where z itself may depend on n.

This same discussion applies to $L_n(z)$ as well, because it is also an nth power. In fact, one does not have to treat both $L_n(z)$ and $H_n(z)$ separately in view of the relation

(5.12) $\max(-X_1, -X_2, \ldots, -X_n) = -\min(X_1, X_2, \ldots, X_n)$

This evidently implies that any asymptotic result on $H_n(z)$ can be transformed to one on $L_n(z)$. Let us therefore concentrate on Z_n and $H_n(z)$.

Note that if $F(z) < 1$ for all z, then $Z_n \to +\infty$ with n (with probability 1). That is, the opposite to it would be that for some finite number A, $Z_n \le A$ for infinitely many n. However, since $Z_n \le Z_{n+1}$, which is evident, $\{Z_n \le A\}$ implies that $\{Z_{n-1} \le A\}$; consequently, if $\{Z_n \le A\}$ for infinitely many n, the same holds for all n. But $P(Z_n \le A) = F^n(A)$, from which

$$P(Z_n \le A \text{ for all } n) \le F^N(A) \qquad \text{for arbitrary } N$$

By letting $N \to +\infty$, we indeed get $P(Z_n \le A$ for all $n) = 0$. Now, if $Z_n \to +\infty$, it cannot have a limiting distribution without normalization. Guided by the central limit theorem, we seek normalization in the form $(Z_n - a_n)/b_n$, where a_n and $b_n > 0$ are two sequences of constants, and look for the possible limiting distribution functions of these normalized maxima. We quote only one theorem on this topic, without proof, and discuss a few examples with specific distribution functions.

In the statement that follows we speak of distribution functions of the same type. Two distribution functions $G(x)$ and $G^*(x)$ are said to be of the same type if there exist constants C and $D > 0$ such that $G(x) = G^*(C + Dx)$.

Theorem 5.4   There are only three types of distribution function that can occur as the limit of

$$P\left(\frac{Z_n - a_n}{b_n} \le z\right) = P(Z_n \le a_n + b_n z) = H_n(a_n + b_n z) = F^n(a_n + b_n z)$$

where a_n and $b_n > 0$ are suitable constants. These three types are represented by the following distribution functions.

(i) $H_{1,\gamma}(z) = \begin{cases} \exp(-z^{-\gamma}) & \text{if } z > 0 \\ 0 & \text{otherwise} \end{cases}$

(ii) $H_{2,\gamma}(z) = \begin{cases} 1 & \text{if } z > 0 \\ \exp(-(-z)^\gamma) & \text{if } z \le 0 \end{cases}$

(iii) $H_{3,0}(z) = \exp(-e^{-z})$ for all z
where $\gamma > 0$ is an arbitrary constant in (i) and (ii).

In the next examples we shall utilize the limit relation

(5.13) $\left(1 + \frac{u_n}{n}\right)^n \to e^u$ whenever $u_n \to u$ as $n \to +\infty$

This can be found in most textbooks on calculus.

Example 5.9   Let $X_1, X_2, \ldots, X_n$ be independent and identically distributed exponential variables with $E(X_j) = 1$. Then the asymptotic distribution of $Z_n - \log n$ is $H_{3,0}(z)$.

Indeed, since $F(x) = 1 - e^{-x}$, $x > 0$,

$$P(Z_n \le \log n + z) = F^n(\log n + z) = (1 - e^{-\log n - z})^n$$

$$= \left(1 - \frac{e^{-z}}{n}\right)^n \to \exp(-e^{-z})$$

from (5.13).   ▲

Example 5.10   If $X_1, X_2, \ldots, X_n$ are uniformly distributed over the interval $(0,1)$, the asymptotic distribution of $n(Z_n - 1)$ is $H_{2,1}(z)$.

By definition of uniform distribution, $F(x) = x$ for $0 \le x \le 1$, and thus for $z < 0$,

$$P\left(Z_n \le 1 + \frac{z}{n}\right) = F^n\left(1 + \frac{z}{n}\right) = \left(1 + \frac{z}{n}\right)^n \to e^z \qquad \blacktriangle$$

Before proceeding to the next examples, note that

$$F^n(x) = \{1 - [1 - F(x)]\}^n = \left\{1 - \frac{n[1 - F(x)]}{n}\right\}^n$$

and thus if $x = a_n + b_n z$ is such that $n[1 - F(x)]$ converges to a function $u(z)$, then, from (5.13), the asymptotic distribution of $(Z_n - a_n)/b_n$ is $e^{-u(z)}$.

Example 5.11   Let the common distribution function of the independent random variables $X_1, X_2, \ldots, X_n$ be $F(x) = 1 - x^{-3}$, $x \ge 1$ (Pareto distribution). Then the asymptotic distribution of $Z_n/n^{1/3}$ is $H_{1,3}(z)$.

In view of the remark preceding this example, it suffices to show that

$$n[1 - F(n^{1/3}z)] \to z^{-3}$$

which is evident by a direct substitution into $1 - F(x) = x^{-3}$.   ▲

The distribution functions in Examples 5.9 to 5.11 are typical of what happens in general concerning the three types of limiting distribution occurring in Theorem 5.4. That is, in order to get $H_{1,\gamma}(z)$ as the limiting distribution of $(Z_n - a_n)/b_n$, $F(x)$ must be smaller than 1 for all x, and we can always choose $a_n = 0$. Furthermore, $n[1 - F(b_n z)]$ should be asymptotically a (negative) power of z. Similarly, one can get $H_{2,\gamma}(z)$ only if the X_j are bounded, in which case a_n is always the least upper bound of X_j. Finally, although $H_{3,0}(z)$ can occur both for bounded and unbounded random variables X_j, what makes it different from the previous two cases is that

$n[1 - F(a_n + b_n z)]$ should asymptotically become e^{-z} as opposed to powers of z in the previous cases. We omit the details and proofs of the preceding statements.

Example 5.12 If $X_1, X_2, ..., X_n$ are independent random variables, each with standard normal distribution, then the asymptotic distribution of $(Z_n - a_n)/b_n$ is $H_{3,0}(z)$, where

$$a_n = (2 \log n)^{1/2} - \frac{(1/2)(\log \log n + \log 4\pi)}{(2 \log n)^{1/2}}, \qquad b_n = (2 \log n)^{-1/2}$$

We first show that for the standard normal distribution function, as $x \to +\infty$,

$$(5.14) \qquad\qquad 1 - N(x) \sim \frac{1}{\sqrt{2\pi} x} e^{-x^2/2}$$

We, in fact, establish two inequalities involving $1 - N(x)$. First, by integrating by parts in

$$(2\pi)^{1/2}[1 - N(x)] = \int_x^{+\infty} e^{-t^2/2}\, dt = \int_x^{+\infty} (te^{-t^2/2}) t^{-1}\, dt$$

we get for $x > 0$,

$$(5.15) \qquad (2\pi)^{1/2}[1 - N(x)] - x^{-1} e^{-x^2/2} = -\int_x^{+\infty} e^{-t^2/2} t^{-2}\, dt$$

Because the right-hand side is negative, (5.15) implies that

$$(5.16) \qquad\qquad 1 - N(x) < (2\pi)^{-1/2} \frac{1}{x} e^{-x^2/2} \qquad x > 0$$

Next, going back to (5.15), and integrating by parts once more, we arrive at the new inequality

$$(5.17) \qquad\qquad 1 - N(x) > (2\pi)^{-1/2} \left(\frac{1}{x} - \frac{1}{x^3} \right) e^{-x^2/2} \qquad x > 0$$

The inequalities (5.16) and (5.17) evidently imply (5.14). Now, writing

$$\frac{n}{\sqrt{2\pi} x} e^{-x^2/2} = \exp\left[-\frac{1}{2} x^2 - \frac{1}{2} \log(2\pi) - \log x + \log n \right]$$

and substituting $x = a_n + b_n z$ with the actual values of a_n and b_n, routine calculation yields that the exponent on the right-hand side above converges to $-z$. Consequently, in view of (5.14), $n[1 - N(a_n + b_n z)]$ converges to e^{-z}, which suffices for the desired limiting distribution in view of those obtained in the paragraph preceding Example 5.11. ▲

Example 5.13 Let $X_1, X_2, ..., X_n$ be independent standard normal variables, and let $n = 18$ million. Let us determine $P(5 < Z_n < 6)$ and $P(-6 < W_n < -5)$.

With the notation of Example 5.12, for $n = 18$ million, $a_n = 5.318$ and $b_n = 0.173$. Hence (with rounded numbers)

$$P(5 < Z_n < 6) = P(a_n - 1.84b_n < Z_n < a_n + 3.94b_n)$$

which, by Example 5.12, equals $H_{3,0}(3.94) - H_{3,0}(-1.84) = 0.979$.

Because of the symmetry of the standard normal distribution about zero, we have, without any further calculation, $P(-6 < W_n < -5) = 0.979$.

▲

Calculations like these can be utilized for different purposes. One possibility is to detect whether an assumption of normality is correct for a set of data. Whether we have 18 million observations, or many fewer, the idea is the same whatever large set of observations we have: We can predict the range within which all observations should fall. Another conclusion that we can draw from the calculations of Example 5.13 is that even though the normal distribution can, in principle, be assigned only to random variables whose range is unlimited, for all practical purposes, normal distributions can also be used in connection with bounded random variables (see, in this connection, Examples 3.17 and 5.13).

5.4 TWO EXACT MODELS VIA LIMIT THEOREMS

Limit theorems are usually associated with approximations. In the present section we establish two theorems, in each of which the exact distribution of a random variable is determined from one of the limit theorems of the previous sections.

Model 1: Errors in Laboratory Measurements

Assume that we measure the length of a small organ under the microscope. If the exact length is m, and if we measured it L, the error $X = L - m$ is a random variable, because it varies from one experiment to another, and from person to person doing the measuring (even if they have equal expertise). We shall deduce from the central limit theorem that the exact distribution of X is normal if X is due exclusively to the inaccuracy of the instrument (including the human factor).

First note the following property of X. If the experimenter was lucky to place the scale to the organ in such a manner that the zero point is exactly at one end of the organ, then with one reading L is obtained. However, more common is a situation where one integer point on the scale is picked out somewhere in the middle and the length is read in two directions, obtaining L_1 and L_2, from which $L = L_1 + L_2$ is recorded. In the latter way, however, an error was committed twice, once to the left and once to the right in the reading of the length. If the errors are denoted by X_1 and X_2, by the assumption that the errors are due to the inaccuracy of the instrument alone, the following assumptions are reasonable: (i) X_1 and X_2 are independent and identically distributed; (ii) X, X_1, X_2, and $X_1 + X_2$ all have the same type of distribution function (because they represent the same type of error); and (iii) the expectation E and the variance V of X are finite. Without loss of generality, we can assume further that $E = 0$ and $V = 1$, because if not, we would deal with $(X - E)/\sqrt{V}$ rather than with X, which has the mentioned properties.

Now, by (ii), there are constants A and $B > 0$ such that the distribution function of X is the same as that of $A + B(X_1 + X_2)$. Hence, by taking expectation and variance, (i) and (iii) imply that $0 = A + 2E_1B$ and $1 = 2V_1B^2$, where E_1 and V_1 are the common expectation and variance, respectively, of X_1 and X_2. That is, $(X_1 + X_2 - 2E_1)/\sqrt{2V_1}$ is distributed as X. We next show that for every n of the form $n = 2^k$, the distribution of $(X_1 + X_2 + \cdots + X_n - nE_1)/\sqrt{nV_1}$ is exactly the same as the distribution of X, where the X_j are independent random variables with the same distribution as X_1 above. Namely, if we take $n = 4$ terms, we know that both $(X_1 + X_2 - 2E_1)/\sqrt{2V_1}$ and $(X_3 + X_4 - 2E_1)/\sqrt{2V_1}$ are distributed as X, and they are independent. Hence their sum has the same type of distribution as does X. By comparing expectation and variance, we thus get the claimed distributional identity for $n = 4$.

Using the same idea, we can therefore reduce the case of $n = 8$ to $n = 4$, then $n = 16$ to $n = 8$, and so on, proving (by induction) the distributional identity for all $n = 2^k$. Let $n \to +\infty$. Since for every n we got the distribution function of X, the limiting distribution, too, has to be the distribution of X. But the central limit theorem tells us that the limiting distribution of $(X_1 + X_2 + \cdots + X_n - nE_1)/\sqrt{nV_1}$ is normal, so that X is normally distributed.

Model 2: The Strength of a Sheet of Metal

Let the strength S of a sheet of metal be measured by the number of times it can be bent before it breaks (breaking usually means that x-rays detect the break). It is evidently a random value because different sheets will withstand different amount of stress. Let us divide the sheet hypothetically into n equal parts, and let $S_1, S_2, \ldots, S_n$ be the strength of the parts. We assume that (i) the strength is proportional to the size; (ii) the S_j are independent (and identically distributed); and (iii) the sheet breaks at its weakest point. We prove that the distribution function of S is Weibull; that is, with some constants $B > 0$ and $c > 0$, $F(z) = 1 - \exp(-Bz^c)$, $z > 0$.

Let us first translate the assumptions into mathematical formulas: (i) For every n, there is a constant d_n such that the distribution function of $d_n S_j$ is the same as that of S; (iii) $S = \min(S_1, S_2, \ldots, S_n)$; and thus (ii) makes it possible to apply the results from Section 5.3. Now, since the distribution of $\min(S_1, S_2, \ldots, S_n)$ is the same distribution for every n, it must have a limiting distribution, which, incidentally, is the distribution of S. By (5.12) and Theorem 5.4, the distribution of S therefore belongs to one of three possible types. But property (i) is satisfied only by one of these three possibilities, that is by the Weibull distribution specified earlier. [Note that, by (5.12), if $H(z)$ is a limiting distribution function for properly normalized maxima, then $1 - H(-z)$ is a limiting distribution for minima. So the Weibull distribution is the one corresponding to $H_{2,\gamma}(z)$ in Theorem 5.4.]

One might disagree with the assumption of independence in (ii). It is, in fact, not necessary. There are dependent models in the literature for the distribution of strength in which the conclusion is the same Weibull distribution as obtained here.

5.5   SUMS OF INDICATOR VARIABLES

In the present section we investigate sums of indicator variables without the assumption of independence. Let

$$(5.18) \qquad I_j = \begin{cases} 1 & \text{with probability } u_j \\ 0 & \text{with probability } 1 - u_j \end{cases}$$

and set

$$(5.19) \qquad m_n = I_1 + I_2 + \cdots + I_n$$

We introduce

$$(5.20) \qquad S_k = \sum{}^* P(I_{i_1} = I_{i_2} = \cdots = I_{i_k} = 1)$$

where Σ^* signifies summation over all $1 \le i_1 < i_2 < \cdots < i_k \le n$. Note that the number of terms in this sum is $\binom{n}{k}$. Therefore, if

$$(5.21) \qquad p_k = P(I_{i_1} = I_{i_2} = \cdots = I_{i_k} = 1)$$

is the same value for all $1 \le i_1 < i_2 < \cdots < i_k \le n$, then

$$(5.22) \qquad S_k = \binom{n}{k} p_k$$

Recall that we have represented both the binomial and the hypergeometric variables in the form (5.19). In both cases (5.21) and (5.22) apply $[p_k = p^k$ in the case of binomial variables, and $p_k = M(M-1)\cdots(M-k+1)/T(T-1)\cdots(T-k+1)$ in the hypergeometric case].

We first establish two inequalities involving the S_k and the distribution of m_n, from which a Poisson limit theorem will be deduced.

Lemma (The Jordan Inequalities) For every integer $s \geq 0$,

$$(5.23) \qquad \sum_{t=0}^{2s+1} (-1)^t \binom{t+r}{r} S_{t+r} \leq P(m_n = r) \leq \sum_{t=0}^{2s} (-1)^t \binom{t+r}{r} S_{t+r}$$

where $S_0 = 1$. The inequalities become an identity if $2s + r \geq n$.

Proof: Setting $J(i_1, i_2, \ldots, i_k)$ for the indicator variable of $I_{i_1} = I_{i_2} = \cdots = I_{i_k} = 1$, we have

$$E(J(i_1, i_2, \ldots, i_k)) = P(I_{i_1} = I_{i_2} = \cdots = I_{i_k} = 1)$$

and thus

$$S_k = E\left[\sum{}^* J(i_1, i_2, \ldots, i_k) \right]$$

On the other hand,

$$\sum{}^* J(i_1, i_2, \ldots, i_k) = \binom{m_n}{k}$$

because both sides count the number of ways of picking k out of m_n. These now yield

$$(5.24) \qquad S_k = E\left[\binom{m_n}{k} \right]$$

(From this formula, the S_k are called the *binomial moments* of m_n.) Hence

$$\sum_{t=0}^{b} (-1)^t \binom{t+r}{r} S_{t+r} = E\left[\sum_{t=0}^{b} (-1)^t \binom{t+r}{r} \binom{m_n}{t+r} \right]$$

where we interchanged summation and taking expectation. Utilizing

$$\binom{t+r}{r} \binom{m_n}{t+r} = \frac{(t+r)!}{t!r!} \frac{m_n!}{(t+r)!(m_n - t - r)!}$$

$$= \frac{m_n!}{r!(m_n-r)!} \frac{(m_n-r)!}{t!(m_n-r-t)!} = \binom{m_n}{r}\binom{m_n-r}{t}$$

where the first factor on the right-hand side does not depend on t, so that it can be removed from behind the summation sign, we get

$$\sum_{t=0}^{b} (-1)^t \binom{t+r}{r} S_{t+r} = E\left[\binom{m_n}{r} \sum_{t=0}^{b} (-1)^t \binom{m_n-r}{t} \right]$$

Note that the expression behind the expectation sign is zero if $m_n < r$, and it is 1 if $m_n = r$. On the other hand, we know (Exercise 26 of Chapter 1) that for $m_n > r$,

$$\sum_{t=0}^{b} (-1)^t \binom{m_n - r}{t} = (-1)^b \binom{m_n - r - 1}{b}$$

Hence, by taking expectations, we get (5.23).   ▲

Theorem 5.5   If, for every fixed k, as $n \to +\infty$, $S_k \to \lambda^k/k!$ for some $\lambda > 0$, then

$$\lim P(m_n = r) = \frac{\lambda^r e^{-\lambda}}{r!} \qquad r = 0, 1, 2, \ldots$$

   Proof: Let $s > 0$ be a fixed number. Then, from the lemma, as $n \to +\infty$,

$$\sum_{t=0}^{2s+1} (-1)^t \binom{t+r}{r} \frac{\lambda^{t+r}}{(t+r)!} \leq \liminf P(m_n = r)$$

$$\leq \limsup P(m_n = r) \leq \sum_{t=0}^{2s} (-1)^t \binom{t+r}{r} \frac{\lambda^{t+r}}{(t+r)!}$$

Letting $s \to +\infty$ yields the same bounds on the two sides, so the limit in the middle exists and it equals

$$\sum_{t=0}^{+\infty} (-1)^t \binom{t+r}{r} \frac{\lambda^{t+r}}{(t+r)!} = \frac{\lambda^r}{r!} \sum_{t=0}^{+\infty} (-1)^t \frac{\lambda^t}{t!}$$

in which we can recognize the Taylor expansion of $e^{-\lambda}$. This completes the proof.   ▲

   As a corollary to Theorem 5.5, the Poisson approximation to the binomial distribution is immediate. Since in this case (5.21) and (5.22) apply with $p_k = p^k$, then if $np \to \lambda$, S_k indeed converges to $\lambda^k/k!$, which is the sole assumption in Theorem 5.5.

   With the same simplicity as above we also get that the hypergeometric distribution (whose parameters are T, M, and t) is asymptotically Poisson if $tM/T \to \lambda > 0$. Namely, for fixed k, (5.21) and (5.22) again imply that $S_k \to \lambda^k/k!$.

   Out of the many possible further applications of Theorem 5.5, we discuss the following classical problem, which has found numerous applications in science and engineering.

   Assume that b balls are distributed into n urns, independently of each other, and that each ball is placed into any one of the urns with the same probability $1/n$. Let us determine the distribution of the number $m_n = m_n(b)$ of those urns which remain empty.

Putting $I_j = 1$ if the jth urn remains empty, and $I_j = 0$ otherwise, we have

$$m_n = I_1 + I_2 + \cdots + I_n$$

Although the I_j are strongly dependent, the evaluation of S_k is easy. Thus $I_{i_1} = I_{i_2} = \cdots = I_{i_k} = 1$ means that all b balls were placed into the $n - k$ urns not identified here by the subscripts. Since the probability of a ball's falling into a specific urn is $1/n$, the probability that it falls into any one of $n - k$ urns is $(n - k)/n$. Furthermore, the balls are placed into the urns independently of each other, and thus

$$P(I_{i_1} = I_{i_2} = \cdots = I_{i_k} = 1) = \left(\frac{n-k}{n}\right)^b = \left(1 - \frac{k}{n}\right)^b$$

That is, (5.21), and thus (5.22) as well, applies, yielding

$$(5.25) \qquad\qquad S_k = \binom{n}{k}\left(1 - \frac{k}{n}\right)^b$$

With these values, the inequalities (5.23) are applicable to the computation of $P(m_n = r)$. Note in the next example that with very moderate values of s, the two inequalities of (5.23) give the value of $P(m_n = r)$ accurate to three decimal digits.

Example 5.14   With the preceding notations, let us compute $P(m_{10}(17) = 3)$ (i.e., that three urns remain empty when 17 balls are placed into 10 urns).

We apply the inequalities (5.23), where $r = 3$ and S_k is given in (5.25) with $n = 10$ and $b = 17$. We get:

$$s = 0: \qquad\qquad S_3 - 4S_4 \le P(m_{10}(17) = 3) \le S_3$$

$$s = 1: \quad S_3 - 4S_4 + 10S_5 - 20S_6 \le P(m_{10}(17) = 3) \le S_3 - 4S_4 + 10S_5$$

where, by (5.25),

$$S_3 = \binom{10}{3}(1 - 0.3)^{17} = 0.27916; \qquad S_4 = \binom{10}{4}(1 - 0.4)^{17} = 0.03555$$

$$S_5 = \binom{10}{5}(1 - 0.5)^{17} = 0.00192; \qquad S_6 = \binom{10}{6}(1 - 0.6)^{17} = 0.000036$$

These values, when substituted into the inequalities above, yield

$$s = 0: \qquad\qquad 0.13697 \le P(m_{10}(17) = 3) \le 0.27916$$

$$s = 1: \qquad\qquad 0.15547 \le P(m_{10}(17) = 3) \le 0.15620$$

from which the first two digits are clearly 0.15, and with rounding in the third digit, we can conclude that $P(m_{10}(17) = 3) = 0.156$.   ▲

Let us now establish a Poisson approximation to the distribution of $m_n(b)$ as a corollary to Theorem 5.5. We once again use the symbol $o(n)$, which signifies a quantity tending to zero when divided by n.

Corollary to Theorem 5.5   In the model of distributing b balls into n urns described previously, if with some number $\lambda > 0$,

$$b = n \log n - n \log \lambda + o(n)$$

then, as $n \to +\infty$,

$$\lim P(m_n(b) = r) = \frac{\lambda^r e^{-\lambda}}{r!} \qquad r = 0, 1, 2, \dots$$

Proof: In view of Theorem 5.5, it suffices to prove that under the conditions above, S_k of (5.25) converges to $\lambda^k/k!$ for every fixed k. When k is fixed, we can replace the binomial coefficient $\binom{n}{k}$ by the asymptotic formula $n^k/k!$. Hence we have to show that

(5.26)
$$n^k \left(1 - \frac{k}{n}\right)^b \sim \lambda^k$$

Taking the logarithm on the left-hand side and utilizing the fact that $\log(1 - z) \sim -z$ as $z \to 0$, we get

$$k \log n + b \log \left(1 - \frac{k}{n}\right) \sim k \log n - b \frac{k}{n}$$

which, by the specific value of b given among the assumptions, becomes $k \log \lambda + o(1)$, which is exactly the logarithmic form of (5.26). The corollary is established.   ▲

Example 5.15   Let us apply the Poisson approximation of the corollary to evaluate the probability that 3 urns remain empty when 300 balls are distributed into 100 urns according to the model of the present discussion.

To evaluate $\log \lambda$, we ignore the $o(n)$ term and set $\log \lambda = \log 100 - 3 = 1.605$, i.e., $\lambda = 4.9787$. Hence

$$P(m_{100}(300) = 3) = \frac{4.9787^3 e^{-4.9787}}{3!} = 0.14157$$

Having computed λ, we could, of course, evaluate

$$P(m_{100}(300) = r)$$

for every r.   ▲

5.6   EXERCISES

1. Let X be a discrete random variable taking on the three values $-1, 1$, and 2 with distribution $(0.2, 0.4, 0.4)$. Let $X_1, X_2, \ldots, X_{81}$ be independent copies of X. Use the central limit theorem to approximate $P(65 < S_{81} < 95)$, where $S_{81} = X_1 + X_2 + \cdots + X_{81}$.

2. If a fair die is rolled 100 times, what is the approximate value of the probability that the sum of the numbers obtained exceeds 400?

3. Pick 100 points from the interval $(0,1)$ independently of one another. Find the (approximate) probability that their average lies between 0.45 and 0.55.

4. Let $X_1, X_2, \ldots, X_{25}$ be independent Poisson variables, each with expected value 0.8. Use the central limit theorem to approximate

$$P(15 < X_1 + X_2 + \cdots + X_{25} < 27)$$

5. Let X_λ be a Poisson variable with parameter λ. Using moment-generating functions, prove that as $\lambda \to +\infty$, $\lim P(X_\lambda - \lambda < \sqrt{\lambda} z) = N(z)$, where $N(z)$ is the standard normal distribution function.

6. Assume that when numbers are rounded off to the nearest integer we in fact add a random variable to these numbers which is uniformly distributed on $(-1/2, 1/2)$. Further assuming that individual round-off errors are independent, find the probability that the sum of 100 rounded numbers will differ by more than 2 from the actual sum of the original numbers.

7. Pick a point x at random in the interval $(0,1)$. Let X_j be the jth digit in the decimal expansion of x. Show that the X_j are independent and identically distributed random variables. By an appeal to the central limit theorem, find (i) $P(400 < X_1 + X_2 + \cdots + X_{100} < 500)$, and (ii) $P(8 < f_j \le 14)$, where f_j is the frequency of the number j, $0 \le j \le 9$, among the digits $X_1, X_2, \ldots, X_{100}$.

8. Let the distribution function of X be standard Cauchy [i.e., its density function $f(z) = c/(1 + z^2)$]. Can the central limit theorem be applied to the sum of independent copies of X?

9. Compute the moment-generating function of a uniformly distributed random variable. By differentiation, find the first three moments of the random variable.

10. Let X be a Poisson variable with $E(X) = 1/2$. Give the moment-generating function of $3X - 6$.

11. Find the moment-generating function of an indicator variable. After decomposing a binomial variable as the sum of independent indicator variables, deduce the form of the moment-generating function of a binomial variable from that of an indicator variable.

12. Let X and Y be independent gamma variables with the same λ-parameter. Show that $X + Y$ also is a gamma variable. (Apply Theorem 3.9.)

13. The gamma function $\Gamma(z)$ is defined by the integral

$$\Gamma(z) = \int_0^{+\infty} x^{z-1} e^{-x} \, dx$$

What is the value of $\Gamma(1/2)$? [Hint: Substitute $u = (2x)^{1/2}$, and refer to the normal distribution function.]

14. Let $X_1, X_2, \ldots, X_{50}$ be independent exponential variables, each with parameter $\lambda = 1/2$. Let Z_{50} be the largest of the X_j, $1 \leq j \leq 50$. Compare the values $P(Z_{50} - 2 \log 50 \leq 3)$ and $\exp(-e^{-1.5})$.

15. Assume that the random life length of a terminally ill patient is exponential with expected value of half a year. Can a hospital claim that their transplant operation on such patients prolongs life because they operated on 50 patients and several have remained alive after two years?

16. Use the inequality (5.16) to estimate $1 - N(5)$.

17. Use the asymptotic theory of extremes to determine u such that $P(Z_{1000} \leq u) \geq 0.99$, where Z_n is the largest of n independent normal variables with common expectation 2 and variance 1/4.

18. Show that if $H(z)$ is the distribution function of the maximum of a set of random variables, then $L(z) = 1 - H(-z)$ is the distribution function of the minimum of some set of random variables, whenever $H(z)$ is continuous.

19. Put $I(C)$ for the indicator variable of the event C. Show that for arbitrary events A and B, $I(A^c) = 1 - I(A)$, $I(A \cap B) = I(A)I(B)$, and $I(A \cup B) = I(A) + I(B) - I(A)I(B)$.

20. Let A, B, and C be arbitrary events. Show, by using indicator variables and taking expectation,

$$S_1 - S_2 \leq P(A \cup B \cup C) \leq S_1 - \frac{2}{3}S_2$$

where $S_1 = P(A) + P(B) + P(C)$ and $S_2 = P(A \cap B) + P(A \cap C) + P(B \cap C)$.

21. With the notation of Section 5.5, show that if as $n \to +\infty$, $S_k \to a^k$ with some $0 < a < 1$, $k \geq 1$, then

$$\lim_{n = +\infty} P(m_n = r) = a^r/(1 + a)^{r+1} \qquad r = 0, 1, 2, \ldots$$

6

Miscellaneous Topics

This chapter contains a few unrelated results from probability theory which were selected for inclusion here either by their surprising conclusion or for their mathematical elegance. Of course, both space and the introductory level put restrictions on the chapter.

Several sections are about random variables which are not necessarily independent. The computation of probabilities of events involving these random variables is sometimes convenient by turning to conditional probabilities. At (3.32) we defined conditioning with respect to a random variable X, even though $X = z$ has probability zero. Because formulas of Chapter 3 in this connection come up several times in the present chapter, we restate some of them, such as (3.33) and (3.34), in the following unified form: Let A be an event defined in terms of inequalities among random variables, including a random variable X whose density function $f(z)$ exists. Then

(6.1) $$P(A) = \int_{-\infty}^{+\infty} P(A|X = z)f(z) \, dz$$

We now turn to specific topics.

6.1 WAITING-TIME PARADOXES

Paradox 1 (Waiting for the Bus) Let $T_1 < T_2 < \cdots < T_k < \cdots$ be the arrival times of buses at stop B. The interarrival times $X_k = T_k - T_{k-1}, k \geq 1$, where, for uniformity of notation, we put $T_0 = 0$, are assumed to be independent and identically distributed exponential variables with $E(X_k) = 15$ (minutes). So, they come quite regularly, except the one that we want to board. Namely, whenever we arrive at stop B, the bus that comes next has longer expectation than 15 minutes from the time the last bus left stop B before our arrival.

For the actual computations for justifying the statement above, we need the fact (see Section 5.1, or one can compute it directly from Theorem 3.9) that, for $k \geq 1$,

$$T_k = X_k + X_{k-1} + \cdots + X_1$$

has a gamma density of the form

(6.2) $$f_k(z) = \frac{1}{(k-1)!}\left(\frac{1}{15}\right)^k z^{k-1} e^{-(1/15)z}, \qquad z > 0$$

Now, let t be our arrival time at stop B, and let $N = N(t)$ be the number of buses that passed through stop B before our arrival (i.e., $T_N < t \leq T_{N+1}$). Furthermore, let $a(t) = t - T_N$ and $r(t) = T_{N+1} - t$. Thus $a(t) > 0$ and

$$T_{N+1} - T_N = a(t) + r(t)$$

implying that

$$E(T_{N+1} - T_N) = E[a(t)] + E[r(t)]$$

Because $a(t) > 0$, $E[a(t)] > 0$ as well. We now show that $E[r(t)] = 15$, establishing the claim that $E(T_{N+1} - T_N) > 15$.

Theorem 6.1   The random variable $r(t)$ is exponentially distributed with $E[r(t)] = 15$.

Proof: By the total probability rule,

(6.3) $$P(r(t) \geq x) = \sum_{n=0}^{+\infty} P(r(t) \geq x \mid N = n)P(N = n)$$

$$= \sum_{n=0}^{+\infty} P(r(t) \geq x, N = n)$$

$$= \sum_{n=0}^{+\infty} P(T_n < t, T_{n+1} \geq t + x)$$

To evaluate the last term, we condition on $T_n = y$, and apply (6.1). Note that given $T_n = y$, the event $\{T_n < t, T_{n+1} \geq t + x\} = \{X_{n+1} \geq t + x - y\}$ if $y < t$, and impossible if $y > t$. Hence, by (6.1) and (6.2), for $n \geq 1$,

$$P(T_n < t, T_{n+1} \geq t + x) = \int_0^t P(X_{n+1} \geq t + x - y)f_n(y)\, dy$$

$$= \int_0^t e^{-b(t+x-y)} \frac{b^n y^{n-1} e^{-by}}{(n-1)!} dy$$

where we put $b = 1/15$. For $n = 0$, $P(T_0 < t, \ T_1 \geq t + x) = P(X_1 \geq t + x) = \exp[-b(t+x)]$. Writing these back into (6.3), and interchanging summation and integration, we get

$$P(r(t) \geq x) = e^{-b(t+x)} + e^{-b(t+x)} \int_0^t \left[\sum_{n=1}^{+\infty} \frac{b^n y^{n-1}}{(n-1)!} \right] dy$$

$$= e^{-b(t+x)} + e^{-b(t+x)} (e^{bt} - 1) = e^{-bx}$$

where we utilized the frequently applied Taylor formula

$$\sum_{n=1}^{+\infty} \frac{b^{n-1} y^{n-1}}{(n-1)!} = e^{by}$$

The theorem is established.   ▲

After having gone through the proof, it is quite easy to explain what happened: because the time t of going to the bus stop was chosen by us, regardless of whether the buses came on time or not on that particular day, we had a small probability of catching the first bus, another small probability of catching the second, and so on, and these probabilities added up to the particular distribution of $r(t)$ that we obtained. (The fact that this distribution is the same as the common distribution of the X_k is a characteristic property of the exponential distribution.)

Paradox 2 (Waiting for New Records) Assume that the water level of river C at city Z is a random variable X with a continuous distribution function $F(z)$. We further assume that dates are chosen so that observations $X_1, X_2, \ldots$ taken on X on these dates are independent (and each has the same distribution as X). Two persons are hired by city Z to measure $X_1, X_2, \ldots$ (at the same time and at the same point, so they should get the same sequence $X_1, X_2, \ldots$, except that they do not know about each other's assignment). The contract for person 1 says that his job is completed as soon as he comes up with an X_j larger than X_1. On the other hand, person 2 completes his job as soon as he gets a value smaller than X_1. Because we assumed $F(z)$ to be continuous, it has probability zero that two X_j be equal, so one of the two persons will complete the job with the second measurement. However, when we ask separately how long it will take for person 1, and for person 2, to complete the job, the answer is the same to each of them: very long. In fact, so long that the expected number of times they have to take measurements is infinity. This paradoxical situation is due to the fact that we do not know in advance which of them will be so lucky as to finish the job in just two steps.

In a bit drier (but more accurate) form, the result is as follows.

Theorem 6.2   Let $X_1, X_2, \ldots$ be independent and identically distributed random variables with continuous distribution function $F(z)$. Let us define two positive integer-valued random variables L and N as follows:

$$L = \text{first } j \quad \text{such that} \quad X_j < X_1$$

$$N = \text{first } j \quad \text{such that} \quad X_j > X_1$$

Then

$$P(L = k) = P(N = k) = \frac{1}{k(k-1)} \qquad k \geq 2$$

implying that

$$E(L) = E(N) = +\infty$$

The random variables X_L and X_N are known in the literature as the (first) lower and upper records, respectively. The indices L and N, therefore, express the waiting times to reach these records.

Proof: First note that L and N do not depend on $F(z)$. They depend only on whether $X_j < X_1$, or $X_j \geq X_1$, and thus instead of the sequence $\{X_j\}$, we can consider $\{F(X_j)\}$, which have the same inequalities as the $\{X_j\}$ have, except perhaps that strict inequalities might become equations. However, by the assumption of the continuity of $F(z)$, equations occur only with probability zero. Now, recall Theorem 3.5, in which it is proved that $F(X_j)$ is uniformly distributed over the interval $(0,1)$. In other words, we established that in regard to L and N, an arbitrary (continuous) $F(z)$ can always be replaced by the uniform distribution. Consequently, we assume from now on that $F(z) = z$, $0 \leq z \leq 1$.

We again use the method of conditioning with respect to X_1. Given $X_1 = z$, $L = k$, $k \geq 2$, if $X_2 \geq z$, $X_3 \geq z, \ldots, X_{k-1} \geq z$ and $X_k < z$. Hence, by the assumption of independence,

$$P(L = k \mid X_1 = z) = (1 - z)^{k-2} z \qquad 0 \leq z \leq 1$$

from which, by (6.1),

$$P(L = k) = \int_0^1 P(L = k \mid X_1 = z) \, dz = \int_0^1 (1 - z)^{k-2} z \, dz$$

This can easily be computed by the substitution $1 - z = u$. We get

$$P(L = k) = \int_0^1 u^{k-2}(1 - u)\, du = \frac{1}{k-1} - \frac{1}{k} = \frac{1}{k(k-1)}$$

Because the distributions of L and N do not depend on $F(z)$, these distributions are necessarily equal. That is, if $L = k$ in the sequence $X_1, X_2, \ldots$, where the X_j are uniformly distributed over $(0,1)$, then $N = k$ in the sequence $(1 - X_1), (1 - X_2), \ldots$ (which also happens to be a uniformly distributed sequence).

Finally,

$$E(L) = \sum_{k=2}^{+\infty} k \frac{1}{k(k-1)} = \sum_{k=2}^{+\infty} \frac{1}{k-1} = +\infty$$

which completes the proof.   ▲

Although we did not deal with infinite expectation in the present book, we can easily deduce from the weak law of large numbers that it implies arbitrarily large values for the averages. Thus, limiting the discussion to L, where $L > 0$ and $E(L) = +\infty$, we can estimate the arithmetical mean of independent copies $L_1, L_2, \ldots, L_s$ of L by the following *truncation method*. Let K be an arbitrary number, and choose M such that

$$\sum_{k=2}^{M-1} k \frac{1}{k(k-1)} \geq K$$

Define the new random variables L_j^* by

$$L_j^* = \begin{cases} L_j & \text{if } L_j < M \\ M & \text{if } L_j \geq M \end{cases}$$

Then $L_j^* \leq L_j$, and

$$E(L_j^*) = \sum_{k=2}^{M-1} k \frac{1}{k(k-1)} + MP(L_j^* = M) > K$$

Hence

$$\frac{L_1 + L_2 + \cdots + L_s}{s} \geq \frac{L_1^* + L_2^* + \cdots + L_s^*}{s}$$

the latter of which is arbitrarily close to their common expectation $E(L_j^*)$ (with probability close to 1), which in turn is larger than K. Because K is arbitrary, we proved that the arithmetical means of the L_j are unbounded.

Paradox 3   (Waiting at a Bank)   Assume that John walks into a bank that has two windows open. At the first window a customer is just stepping up for service, and at window 2 one customer is being served, but otherwise the number of customers in each line is the same at John's arrival. Does John have an advantage by joining line 2 if the service times are independent exponential variables?

The answer is no; and after some reflection one recognizes that the mathematics behind this problem is the same as in the case of waiting for the bus. That is, in one line, the required service time before John would be served is $X_1 + X_2 + \cdots + X_n$, where the X_j are independent and identically distributed random variables, and n is the number of customers in the line (window 1). At window 2, n is the same but only $n - 1$ customers require full service and one customer "fractional" service, which is the same as $r(t)$ in the case of the bus paradox [i.e., before John's turn comes at window 2, $X_1 + X_2 + \cdots + X_{n-1} + r(t)$ is the waiting time]. But since in both models, the underlying distribution is exponential, we can refer to the bus paradox to get that the distribution of $r(t)$ is the same as that of X_j, so the two sums have identical expectations.

6.2   AN UNEXPECTED SITUATION CONCERNING INDEPENDENT RANDOM VARIABLES

What we describe in this section is again valid for arbitrary random variables with continuous distribution function. However, since the discussion involves inequalities only, we can transform the random variables to the uniform distribution by an appeal to Theorem 3.5 (just as in the case of the second paradox of the preceding section).

Let X, Y, and Z be independent random variables, uniformly distributed over the interval $(0,1)$. Let us call urn 1 the interval $(0,X)$ and urn 2 the inerval $(X,1)$. Now, the probability that Y or Z would fall into urn 1 is 1/2, because "Y is in urn 1" means that $Y < X$, and since $P(Y = X) = 0$, by symmetry, $P(X < Y) = P(X > Y) = 1/2$. If the reader is more convinced by an analytic argument, condition on X, and apply (6.1):

$$P(X > Y) = \int_0^1 P(X > Y \mid X = z)\, dz = \int_0^1 P(Y < z)\, dz = \int_0^1 z\, dz = \frac{1}{2}$$

So X, Y, and Z are independent, urns 1 and 2 are defined by X alone, and

$$P(Y \text{ falls into urn } 1) = P(Z \text{ falls into urn } 1) = \frac{1}{2}$$

Is it true that $P(\text{both } Y \text{ and } Z \text{ fall into urn } 1) = \frac{1}{4}$?
The answer is no. Namely,

$$\{\text{both } Y \text{ and } Z \text{ fall into urn } 1\} = \{\max(Y,Z) < X\}$$

whose probability, evaluated by the analytic method, equals

$$P(\max(Y,Z) < X) = \int_0^1 P(\max(Y,Z) < X \,|\, X = z) \, dz$$

$$= \int_0^1 P(Y < z, \, Z < z) \, dz = \int_0^1 z^2 \, dz = \frac{1}{3}$$

It is instructive to evaluate this same probability by the combinatorial argument as above, referring to symmetry. Because of symmetry, there are six ways of placing the points X, Y, and Z on the interval $(0,1)$. Of these, only two, (Y,Z,X) and (Z,Y,X), favor the event that both Y and Z fall below X. So the desired probability is $2/6 = 1/3$.

Let us extend this investigation to several random variables. Let $X_1, X_2, ..., X_n$ be independent random variables, each with uniform distribution on the interval $(0,1)$. We denote the smallest, the second smallest, ..., the rth smallest, ..., by $X_{1:n} \le X_{2:n} \le \cdots \le X_{r:n} \le \cdots \le X_{n:n}$. Because of the continuity of the uniform distribution function, equations occur only with probability zero. Therefore, we have, with probability 1, $n + 1$ intervals $(X_{j:n}, X_{j+1:n})$, $0 \le j \le n$, where $X_{0:n} = 0$ and $X_{n+1:n} = 1$. We call these intervals urns 1 through $n + 1$. If we now start throwing additional uniformly distributed random variables $Y_1, Y_2, ...$ into the interval $(0,1)$, they fall into the urns with equal probabilities; that is, the probability that Y_j falls into urn k is $1/(n + 1)$, regardless of the value of k.

The proof of this claim with the combinatorial argument is as simple as in the case of $n = 1$. Namely, when a Y is also placed into $(0,1)$, we have $n + 1$ points, which, by symmetry, have $(n + 1)!$ possible ways of having been placed there. Now, if Y is in the kth urn, Y has a fixed position among the X_j, but the X_j can freely be interchanged without disturbing the label of the urn in which Y is. That is, the number of favorable cases to Y's being in urn k is $n!$, yielding $n!/(n + 1)! = 1/(n + 1)$ for its probability. Once again, however, if several independent Y's are placed into the urns, their placements are not independent. As a matter of fact, if we repeat the argument above, we get that the probability of both Y_1 and Y_2 being in urn k is $2(n!)/(n + 2)! = 2/(n + 1)(n + 2)$. (Note that Y_1 and Y_2 can be interchanged between themselves when counting the number of favorable cases.)

With one more reference to symmetry, we get that every possible placement of the independent and uniformly distributed $Y_1, Y_2, ..., Y_m$ into "urns" 1 through $n + 1$ has the same probability, leading to the Bose–Einstein statistic introduced in Section 1.9. Since this model of placing the Y's among the X's has a well-developed theory in statistics (under the strange name "number of exceedances"), those results are directly applicable to physics, and vice versa. This relation alone justifies the inclusion in this chapter of the model described above. The main motive, however, was the demonstration of the "loss of independence of the independent random variables Y_j" as they were placed among the X_j.

6.3   ONCE MORE ON INDEPENDENCE

Let X and Y be independent exponential variables with $E(X) = E(Y) = 1$. Put $W = \min(X,Y)$ and $R = |X - Y|$. Although both W and R are computed from both X and Y, W and R are independent. An additional "surprise" is that the distribution of R is also exponential with $E(R) = 1$; that is, R is "as large" as X or Y.

Let us compute the joint density of W and R. That is, we have to find, for $u > 0$ and $v > 0$,

$$(6.4) \qquad \lim \frac{P(u < W \leq u + \triangle u, v < R \leq v + \triangle v)}{\triangle u \, \triangle v} = f(u,v)$$

as both $\triangle u$ and $\triangle v$ tend to zero. Now if $W = X$, then $R = Y - X$, and if $W = Y$, then $R = X - Y$. Because these two cases are symmetric in X and Y,

$$(6.5) \qquad P(u < W \leq u + \triangle u, \, v < R \leq v + \triangle v) =$$

$$2P(u < X \leq u + \triangle u, \, v < Y - X \leq v + \triangle v)$$

We compute the right-hand side by an appeal to (6.1). We get

$$(6.6) \qquad P(u < X \leq u + \triangle u, \, v < Y - X \leq v + \triangle v)$$

$$= \int_u^{u+\triangle u} P(v < Y - X \leq v + \triangle v \mid X = z)e^{-z} \, dz$$

$$= \int_u^{u+\triangle u} P(z + v < Y \leq z + v + \triangle v)e^{-z} \, dz$$

$$= \int_u^{u+\triangle u} (e^{-z-v} - e^{-z-v-\triangle v})e^{-z} \, dz = e^{-v}(1 - e^{-\triangle v}) \int_u^{u+\triangle u} e^{-2z} \, dz$$

$$= \frac{1}{2}e^{-v}(1 - e^{-\triangle v})(e^{-2u} - e^{-2u-2\triangle u}) = \frac{1}{2}e^{-v}e^{-2u}(1 - e^{-\triangle v})(1 - e^{-2\triangle u})$$

Next, observe that, either by a Taylor expansion or by L'Hospital's rule,

$$\lim_{\triangle v=0} \frac{1 - e^{-\triangle v}}{\triangle v} = 1 \qquad \text{and} \qquad \lim_{\triangle u=0} \frac{1}{2}\frac{1 - e^{-2\triangle u}}{\triangle u} = 1$$

and thus, by (6.4) to (6.6),

$$f(u,v) = 2e^{-2u}e^{-v} \qquad u > 0, \quad v > 0$$

Evidently, $f(u,v) = 0$ if either u or v is negative. Hence, if we write

$f_1(u) = 2e^{-2u}$, $u > 0$, and $f_2(v) = e^{-v}$, $v > 0$, we obtain that $f(u,v) = f_1(u)f_2(v)$, which is one of the criteria for independence of the marginals W and R. We also see that both marginal densities are exponential, and that $E(W) = 1/2$ and $E(R) = 1$.

The result above can be extended to an arbitrary number of exponential variables. With no essential change in the calculations above, one gets the following property of the exponential distribution.

Let $X_1, X_2, \ldots, X_n$ be independent exponentially distributed random variables with $E(X_j) = 1$ for all j. Let us rearrange the X_j into an increasing sequence $X_{1:n} \le X_{2:n} \le \cdots \le X_{n:n}$. (The new random variables $X_{r:n}$, $1 \le r \le n$, are known as the order statistics of the X_j, $1 \le j \le n$.) Now the extension of the independence of W and R, which with the current notation become $W = X_{1:2}$ and $R = X_{2:2} - X_{1:2}$, states that the differences

$$d_j = X_{j:n} - X_{j-1:n}, \qquad 1 \le j \le n \qquad (X_{0:n} = 0)$$

are independent exponential variables with $E(d_j) = 1/(n - j + 1)$, $1 \le j \le n$. This implies that the gap between the largest $X_{n:n}$ and the second largest $X_{n-1:n}$ is always as large as the X_j themselves. We also reobtained in this statement an earlier observation (Example 4.15) that the minimum $X_{1:n}$ is exponentially distributed with expectation $1/n$.

6.4   THE MEAN AND STANDARD DEVIATION FOR NORMAL SAMPLES

Let $X_1, X_2, \ldots, X_n$ be independent and identically distributed normal variables. Put

$$X_{av} = \frac{1}{n}(X_1 + X_2 + \cdots + X_n)$$

and

$$\sigma_n^2 = \frac{1}{n-1}\sum_{j=1}^{n}(X_j - X_{av})^2$$

In statistics, n independent and identically distributed random variables are called a sample of size n, and X_{av} the mean and σ_n the standard deviation of the sample. One of the main reasons why statistics, based on normal samples, is so successful and appealing is the fact that X_{av} and σ_n^2 are independent random variables. This presents a surprise similar to the independence part of the result of the preceding section, because both X_{av} and σ_n^2 are functions of the same X_j. Therefore, both the significance of the result, and the surprise contained in it, justify the inclusion of the present section in this chapter. Besides, the proof itself is very instructive and elegant.

We start with some review of matrix algebra. We investigate $n \times n$ matrices $\mathbf{C} = (c_{ij})$ with the following properties: For all $1 \leq i \leq n$ and $1 \leq t \leq n$, $i \neq t$,

$$\sum_{j=1}^{n} c_{ij}^2 = 1 \qquad \text{and} \qquad \sum_{j=1}^{n} c_{ij}c_{tj} = 0$$

Matrices with these two properties are called *orthogonal*. Note that the meaning of the two assumptions is that, if $\mathbf{C}'$ denotes the transpose of $\mathbf{C}$, then $\mathbf{CC}' = \mathbf{I}$, where $\mathbf{I}$ is the $(n \times n)$ unit matrix (i.e., its diagonal entries are 1 and all others are zero). It therefore means that $\mathbf{C}'$ is the inverse of $\mathbf{C}$. But then the inverse of $\mathbf{C}'$ is $\mathbf{C}$ (i.e., $\mathbf{C}'\mathbf{C} = \mathbf{I}$), which in turn means that $\mathbf{C}'$, too, is orthogonal.

Now let $\mathbf{C} = (c_{ij})$ be an orthogonal matrix, and consider the following linear transformations (or system of equations):

(6.7) $\qquad\qquad u_i = c_{i1}z_1 + c_{i2}z_2 + \cdots + c_{in}z_n \qquad 1 \leq i \leq n$

which can be abbreviated as $\mathbf{u} = \mathbf{z}\mathbf{C}'$, where $\mathbf{u} = (u_1, u_2, \ldots, u_n)$ and $\mathbf{z} = (z_1, z_2, \ldots, z_n)$. Applying the transpose sign to a vector when it is written in a column, and viewing vectors as $(1 \times n$ or $n \times 1)$ matrices, we have

(6.8) $\qquad\qquad \sum_{j=1}^{n} u_j^2 = \mathbf{uu}' = \mathbf{zC'Cz}' = \mathbf{zIz}' = \mathbf{zz}' = \sum_{j=1}^{n} z_j^2$

We now return to our original problem of investigating X_{av} and σ_n^2. Note that their dependence or independence is not affected by the common expectation and variance of the X_j, which, therefore, may be assumed to be standard normal. We first prove the following basic result.

Theorem 6.3 Let $X_1, X_2, \ldots, X_n$ be independent standard normal variables, and let $\mathbf{C} = (c_{ij})$ be an orthogonal matrix. Then the random variables

$$Y_i = c_{i1}X_1 + c_{i2}X_2 + \cdots + c_{in}X_n \qquad 1 \leq i \leq n$$

also are independent standard normal variables.

Proof: We apply (3.23), which yields

$$P(Y_i \leq v_i, \, 1 \leq i \leq n) = (2\pi)^{-(1/2)n} \int\int \cdots \int \exp\left(-\frac{1}{2}\sum_{j=1}^{n} z_j^2\right) dz_1 \, dz_2 \cdots dz_n$$

where the integration is over those points $(z_1, z_2, \ldots, z_n)$ for which

$$c_{i1}z_1 + c_{i2}z_2 + \cdots + c_{in}z_n \leq v_i \qquad 1 \leq i \leq n$$

Let us substitute u_i, $1 \le i \le n$, defined at (6.7). Then, since the equation $\mathbf{CC'} = \mathbf{I}$ implies that the determinant of $\mathbf{C}$ is 1, $dz_1 \, dz_2 \cdots dz_n = du_1 \, du_2 \cdots du_n$. Furthermore, the new integration is over $u_i \le v_i$, $1 \le i \le n$, and, by virtue of (6.8), the integrand has the same functional form in the u_i as it is in the z_i. Therefore, the n-fold integral now splits into a product of n integrals, each of which can be recognized to be the standard normal distribution function. But this is exactly the definition of independence, and thus the theorem is established.   ▲

Let now $\mathbf{C}$ be an orthogonal matrix with $c_{1j} = n^{-1/2}$. It is not difficult to see that in an orthogonal matrix one row can freely be chosen, so $\mathbf{C}$ does exist with the choice above. Then, with the notation of Theorem 6.3, $n^{-1/2} Y_1 = X_{av}$. We also rewrite σ_n^2 by means of Y_j. Since, by the elementary calculation,

$$(n-1)\sigma_n^2 = \sum_{j=1}^{n} (X_j^2 - 2X_j X_{av}) + nX_{av}^2 = \sum_{j=1}^{n} X_j^2 - nX_{av}^2$$

(6.8) yields

$$(n-1)\sigma_n^2 = \sum_{j=1}^{n} Y_j^2 - Y_1^2 = \sum_{j=2}^{n} Y_j^2$$

From these new expressions for X_{av} and σ_n^2, and from the independence of the Y_i, guaranteed by Theorem 6.3, the independence of X_{av} and σ_n^2 follows.

6.5   THE BOREL–CANTELLI LEMMA; THE STRONG CONVERGENCE OF THE RELATIVE FREQUENCY

Let A be an event in connection with a random experiment Ω. Let $P(A) = p$, and let $k_A(n)$ denote the frequency of A in n independent repetitions of Ω. We established in Chapter 2 that for every $\varepsilon > 0$,

$$(6.9) \qquad P\left(\left| \frac{k_A(n)}{n} - p \right| \ge \varepsilon \right) \le \frac{p(1-p)}{n\varepsilon^2}$$

This, combined with the more accurate normal approximation when n is large, is sufficient for all practical purposes to conclude that the relative frequency and the probability of the event A "are close to each other" as n increases indefinitely. From a theoretical point of view, however, "being close" is more convincing if an actual limit can be established, the type that we got accustomed to in calculus. Well, probability cannot be avoided completely, but the following result is good enough to be called strong convergence (of the relative frequency).

Theorem 6.4   With the preceding notations,

$$P\left(\lim_{n=+\infty} \frac{k_A(n)}{n} = p\right) = 1$$

We first prove a simple inequality.

Lemma   Let $A_1, A_2, \ldots$ be an infinite sequence of events. Then

$$P\left(\bigcup_{j=1}^{+\infty} A_j\right) \le \sum_{j=1}^{+\infty} P(A_j)$$

Proof: In Chapter 1 we proved a similar inequality for a finite sequence of events. That is, we know (Theorem 1.5) that

$$P\left(\bigcup_{j=1}^{n} A_j\right) \le \sum_{j=1}^{n} P(A_j)$$

in which the right-hand side is further increased if the summation is extended to infinity. However, in the new inequality

$$P\left(\bigcup_{j=1}^{n} A_j\right) \le \sum_{j=1}^{+\infty} P(A_j)$$

the right-hand side does not depend on n; consequently, it remains valid if we let $n \to +\infty$. Because the sequence of unions of events is increasing with n, we can apply the Lemma of Section 3.2, which says that the limit of the left-hand side above is the probability of the infinite union. The Lemma is established.   ▲

The basic tool of proving Theorem 6.4 is the following result, which is known as the *Borel–Cantelli lemma*.

Theorem 6.5   Let $A_1, A_2, \ldots$ be an infinite sequence of events. Assume that

$$\sum_{j=1}^{+\infty} P(A_j) < +\infty$$

Then, with probability 1, only a finite number of the A_j occur at the same time.

Proof: Let B be the event that infinitely many A_j occur. Then, for every N,

$$B \subset \bigcup_{j=N}^{+\infty} A_j$$

Hence, by the monotonicity of probability (Theorem 1.3) and by the

Lemma,

$$P(B) \le P\left(\bigcup_{j=N}^{+\infty} A_j\right) \le \sum_{j=N}^{+\infty} P(A_j)$$

Now, since $P(B)$ does not depend on N, the inequality remains to hold if we let $N \to +\infty$. However, in virtue of the convergence of the infinite sum in the theorem, the limit of the extreme right-hand side, as $N \to +\infty$, is zero, implying that $P(B) = 0$ or $P(B^c) = 1$. This completes the proof. ▲

We can now turn to the proof of Theorem 6.4.

Proof of Theorem 6.4: Let us introduce the indicator variables

$$I_j = \begin{cases} 1 & \text{if } A \text{ occurs at the } j\text{th repetition} \\ 0 & \text{otherwise} \end{cases}$$

Then

(6.10) $$k_A(n) = I_1 + I_2 + \cdots + I_n$$

We first show that, as $n \to +\infty$, with probability 1,

(6.11) $$\lim \frac{k_A(n^2)}{n^2} = p$$

If we define the events

(6.12) $$A_n = \left[\left|\frac{k_A(n^2)}{n^2} - p\right| \ge \varepsilon\right]$$

where $\varepsilon > 0$ is an arbitrary fixed number, then, by (6.9),

$$P(A_n) \le \frac{p(1-p)}{n^2 \varepsilon^2} = \frac{c}{n^2}$$

We thus have

$$\sum_{n=1}^{+\infty} P(A_n) \le c \sum_{n=1}^{+\infty} \frac{1}{n^2} < +\infty$$

implying that Theorem 6.5 is applicable. This yields that the inequality in (6.12) can hold only for a finite number of n; that is, there is an n_0 such that, for all $n \ge n_0$, the reversed inequality holds in (6.12). Since $\varepsilon > 0$ is arbitrary, this means that $k_A(n^2)/n^2 - p$ converges to zero, and thus (6.11) is established (note that the argument above is valid with probability 1 only, because we appealed to Theorem 6.5).

Now let m be an arbitrary positive integer. Let the integer n be such that $n^2 < m \le (n+1)^2$. Then, in view of (6.10),

$$0 < k_A(m) - k_A(n^2) \le k_A((n+1)^2) - k_A(n^2) = I_{n^2+1} + \cdots + I_{(n+1)^2}$$
$$\le (n+1)^2 - n^2 = 2n + 1$$

from which, after dividing by n^2 and applying (6.11), we get that, with probability 1, as m (and thus n as well) tends to $+\infty$,

$$\lim \frac{k_A(m)}{n^2} = p$$

However, since $0 < m - n^2 \le (n+1)^2 - n^2 = 2n + 1$, $m/n^2 \to 1$ as both tend to $+\infty$; thus the limit above is not effected if n^2 is replaced by m in the denominator. Theorem 6.4 is thus established.   ▲

Note that the independence of the repetitions of Ω did not play very much role. It came up through the inequality of (6.9) only, which in turn was utilized in establishing (6.11) through Theorem 6.5. Therefore, the conclusion of Theorem 6.4 remains valid if the variables I_j in (6.10) are such that $E(I_j) = p$ for each j, and $V(k_A(n)/n) \le g(n)$ with

(6.13) $$\sum_{n=1}^{+\infty} g(n^2) < +\infty$$

In this new formulation of Theorem 6.4, even the adjective "indicator" can be dropped in connection with the variables I_j [in which case the subscript A in $k_A(n)$ loses its meaning], because we never used that their only values are zero and 1. Their boundedness was, however, utilized. We therefore proved the following more general result.

Theorem 6.6   Let $I_1, I_2, \ldots, I_n$ be bounded random variables such that $E(I_j) = p$ for each j, and their sum

$$k(n) = I_1 + I_2 + \cdots + I_n$$

satisfies $V(k(n)/n) \le g(n)$, for which (6.13) holds. Then, with probability 1, as $n \to +\infty$,

$$\lim \frac{k(n)}{n} = p$$

If the random variables I_j in Theorem 6.6 are independent and identically distributed, one can take $g(n) = c/n$ with some constant c, so that the conclusion of the theorem is valid.

6.6   FAIR GAMES

Assume that two players A and B play a game in which both have the same chance (hence, 1/2 each) of winning. If they repeat the game a large number of times, and independently of each other, what can we say about the gain of either player?

Counting the profit from the point of view of A, we introduce the quantities

$$g_j = \begin{cases} 1 & \text{if } A \text{ wins the } j\text{th game} \\ -1 & \text{if } B \text{ wins the } j\text{th game} \end{cases}$$

and

$$G_n = g_1 + g_2 + \cdots + g_n$$

Evidently, G_n is the aggregate gain by A in n games. Because we assumed equal chances for the two players, $P(g_j = 1) = P(g_j = -1) = 1/2$, and thus $E(g_j) = 0$, $E(G_n) = 0$, and, by Theorem 6.6, as $n \to +\infty$,

(6.14)
$$\lim \frac{G_n}{n} = 0$$

with probability 1. The games therefore appear to be fair from every point of view. But if fairness means that we cannot win (and cannot lose) in the long run, the games become quite boring. Well, chance took care of that, too. By the end of this section we shall see that fair games are still fair, but are not likely to be boring: We probably win a lot or lose a lot.

First let us review what we already know about G_n. Formula (6.14) is the main reason we call the present sequence of games fair. But what does (6.14) actually say? It says only that G_n is of smaller order of magnitude than n, but it can still be a large positive or a small negative number. The central limit theorem is, in fact, a better guide to the magnitude of G_n. Two facts follow from the central limit theorem, which we apply in the form of (5.3). First, it implies that if we can win a lot, then with the same probability, we can lose the same amount (because both the exact distribution of G_n, and its approximation by the normal distribution, are symmetric about zero). The second implication of (5.3) is that G_n can indeed become as large as $b(nV)^{1/2}$, where b is a fixed number and V is the common variance of the g_j (i.e., $V = 1$). But when n is large, $n^{1/2}$ is relatively small, and thus such a growth for G_n is rather a confirmation of fairness than a fact that could be used against it. For example, if A and B played $n = 10000$ times, then, by (5.3), $P(G_n > 100) = 1 - N(1) = 0.1587$, but $P(G_n \le 250) = N(2.5) = 0.9938$. So G_n can grow beyond 100 but it would rarely get as high as 250 (both in 10,000 games).

Assume now that G_n did grow to a certain amount. Is it guaranteed that A will lose it if they continue playing? Yes, indeed. Thus the following theorem is true when the games are continued indefinitely.

Theorem 6.7   With the notations of the present section, $G_n = 0$ infinitely many times with probability 1.

Proof: Evidently, G_n can be zero only if n is even. Let $n = 2m$. Then $G_{2m} = 0$ means that out of $2m$ repetitions, exactly m times, $g_j = 1$. Hence, by the binomial distribution formula,

(6.15)
$$P(G_{2m} = 0) = \binom{2m}{m}\left(\frac{1}{2}\right)^{2m}$$

Let p_{2k} be the probability that $2k$ is the first time that $G_{2k} = 0$. Then, because of the independence of the g_j, the sequence of the games can be viewed as if all games started at $2k$, independently of the past, and thus

$$P(G_{2m} = 0) = p_{2m} + \sum_{k=1}^{m-1} p_{2k} P(G_{2m-2k} = 0)$$

Let us multiply this equation by z^{2m} and sum over all positive m. We then get the relation

$$U(z) = T(z) + U(z)T(z)$$

where

$$U(z) = \sum_{m=1}^{+\infty} P(G_{2m} = 0)z^{2m}$$

and

$$T(z) = \sum_{m=1}^{+\infty} p_{2m} z^{2m}$$

Because p_{2m}, $m \geq 1$, represent the probabilities of mutually exclusive events, the sum of all p_{2m} [i.e., $T(1)$] is a probability, namely, the probability that G_{2m} ever becomes zero. We therefore want to compute $T(1)$, which we do with the help of $U(z)$. We write our previous relaton between $T(z)$ and $U(z)$ as

(6.16)
$$T(z) = \frac{U(z)}{1 + U(z)}$$

from which we try to find $T(1)$. A direct substitution of $z = 1$ does not help, because, as we shall soon see, $U(1)$ is not finite. However, if we do show that $U(1)$ is infinite, then by letting $z \rightarrow 1$, we get 1 as the limit on the right-hand side, which yields $T(1) = 1$. This then completes the proof of the theorem because, as we observed earlier, if $G_{2m} = 0$ for some m, then everything starts again, so if with probability 1 G_{2m} does become zero, it becomes zero for the second time, then for the third time, and so on, that is, it becomes zero, with probability 1, an arbitrary number of times.

For completing the proof, we thus have to show that $U(1)$ is not finite. Because (6.15) is a binomial probability, we can use asymptotic expressions from Chapter 2. Formulas (2.40) and (2.42) with $p = 1/2$ and $x_m = 0$ yield that, as $m \to +\infty$,

$$P(G_{2m} = 0) \sim Cm^{-1/2}$$

where $C > 0$ is a constant. Hence, from the well-known result of calculus that the sum of $m^{-1/2}$ over all m is divergent, we get $U(1) = +\infty$, which completes the proof.   ▲

While Theorem 6.7 guarantees that the players break even many times in the course of the games, a more practical question is: How many times will A lead in the games if they play n times? We say that A is leading if either $G_k > 0$, or if $G_k = 0$ but $G_{k-1} = 1$. Let us denote by L_n the number of times when A is leading in the course of n games. The following relations are true.

Theorem 6.8   For even integers $n = 2m$,

$$P(L_{2m} = 2k) = \frac{\binom{2k}{k}\binom{2m - 2k}{m - k}}{2^{2m}} \qquad k = 0, 1, \ldots$$

Theorem 6.9   As $n \to +\infty$,

$$\lim P\left(\frac{L_n}{n} \le z\right) = \frac{2}{\pi} \arcsin \sqrt{z} \qquad 0 < z < 1$$

Remark   Writing

$$\arcsin \sqrt{b} - \arcsin \sqrt{a} = \int_a^b \frac{1}{\sqrt{x(1 - x)}} dx$$

which expresses the probability that L_n lies between an and bn, we see a strange result. Because the integrand $[x(1 - x)]^{-1/2}, 0 < x < 1$, is a U-shaped curve with its minimum at $x = 1/2$, for a given length $b - a$, we get the smallest value above when a and b are symmetric around $1/2$. In other words, the least likely case among all possibilities is that L_n is around $(1/2)n$, representing a nearly break-even situation. This is what we promised when we said earlier that the games will not become boring.

Proof of Theorem 6.8:   We prove it by induction over m. For $m = 1, P(L_2 = 0) = P(L_2 = 2) = 1/2$, because $L_2 = 0$ is equivalent to $g_1 = -1$, and $\{L_2 = 2\} = \{g_1 = 1\}$. By a substitution, we find that the formula of the theorem also gives 1/2 for these probabilities.

We now establish a recursive formula by an argument similar to the one in the proof of Theorem 6.7. Let B_t be the event that $2t$ is the first time that $G_{2j} = 0$. Then, applying the total probability rule with the complete system $B_t \cap \{g_1 = 1\}$ and $B_t \cap \{g_1 = -1\}$, $t \geq 1$, we get, in virtue of the independence of the sequence g_j, $j \geq 1$,

$$P(L_{2m} = 2k) = \sum_{t=1}^{m} \frac{1}{2}\left[P(L_{2m-2t} = 2k - 2t) + \right.$$

$$\left. P(L_{2m-2t} = 2k) \right]P(B_t) \qquad 0 < k < m$$

where $P(L_0 = j) = 1$ if $j = 0$, and zero otherwise. Next, we compute $P(B_t) = p_{2t}$, where we used the notation of the preceding proof. From (6.15) we find that

$$U(z) = \frac{1}{\sqrt{1 - z^2}} - 1$$

[simply compute the Taylor expansion of the right-hand side and comp..re the coefficients with (6.15)], which, from (6.16), yields

$$p_{2t} = P(G_{2t-2} = 0) - P(G_{2t} = 0)$$

We can thus compute p_{2t} explicitly by (6.15). Writing these formulas back into the recursive formula for $P(L_{2m} = 2k)$, $0 < k < m$, and assuming that the theorem is true for $P(L_{2m-2t} = 2j)$ for $t \geq 1$ and for all j, we get the formula of the theorem for $P(L_{2m} = 2k)$, $0 < k < m$. The recursive formula is somewhat modified when $k = 0$ or $k = m$. Thus $P(L_{2m} = 2k \mid B_t, g_1 = 1) = 0$ if $0 \leq k < m$ and $t > m$, but these terms will become 1 if $k = m$. Similarly, $P(L_{2m} = 0 \mid B_t, g_1 = -1) = 1$ for all $t > m$, which were zero in the previous recursive formula. But with the addition of these terms, we similarly get the theorem by induction, this time for all $0 \leq k \leq m$, as in the case of $0 < k < m$. The theorem is established.   ▲

Proof of Theorem 6.9:   Because $L_{2m} \leq L_{2m+1} \leq L_{2m} + 1$, it suffices to prove the theorem with even index. We can therefore use the exact formula of Theorem 6.8. Note that the formula of Theorem 6.8 is the product of two binomial probabilities of the type of (6.15). Hence we can again go back to formulas (2.40) and (2.42) (with $p = 1/2$ and $x_m = 0$) to obtain, for any integer k with $ma < k < mz$,

$$P(L_{2m} = 2k) \sim \frac{1}{\pi\sqrt{k(m-k)}} = \frac{1}{\pi} \frac{1}{\sqrt{(k/m)(1 - k/m)}} \frac{1}{m}$$

If we add these terms up for all k with $ma < k < mz$, then, with the notation

$x_k = k/m$ and thus $x_{k+1} - x_k = 1/m$, we recognize a Riemann sum of the integral

$$\frac{1}{\pi} \int_a^z \frac{1}{\sqrt{x(1-x)}} dx$$

from which, by letting $a \to 0$, the theorem follows.   ▲

Index